Our 15-day trial period lets you examine the documentation and return the package – with seal unbroken – for a full refund.

Package not returnable if label is cut or removed.

KNAPSACK PROBLEMS

WILEY–INTERSCIENCE
SERIES IN DISCRETE MATHEMATICS AND OPTIMIZATION

Aarts and Korst SIMULATED ANNEALING AND BOLTZMANN MACHINES: A STOCHASTIC APPROACH TO COMBINATORIAL OPTIMIZATION AND NEURAL COMPUTING

Ahlswede and Wegener SEARCH PROBLEMS

Anderson and Nash LINEAR PROGRAMMING IN INFINITE-DIMENSIONAL SPACES: THEORY AND APPLICATION

Beale INTRODUCTION TO OPTIMIZATION

Fishburn INTERVAL ORDERS AND INTERVAL GRAPHS: A STUDY OF PARTIALLY ORDERED SETS

Gondran and Minoux GRAPHS AND ALGORITHMS
(Translated by S. Vajda)

Goulden and Jackson COMBINATORIAL ENUMERATION

Graham, Rothschild, and Spencer RAMSEY THEORY

Gross and Tucker TOPOLOGICAL GRAPH THEORY

Hall COMBINATORIAL THEORY, SECOND EDITION

Lawler, Lenstra, Rinnooy Kan, and Shmoys, Editors THE TRAVELING SALESMAN PROBLEM: A GUIDED TOUR OF COMBINATORIAL OPTIMIZATION

Martello and Toth KNAPSACK PROBLEMS: ALGORITHMS AND COMPUTER IMPLEMENTATIONS

Minc NONNEGATIVE MATRICES

Minoux MATHEMATICAL PROGRAMMING: THEORY AND ALGORITHMS
(Translated by S. Vajda)

Nemhauser and Wolsey INTEGER AND COMBINATORIAL OPTIMIZATION

Nemirovsky and Yudin PROBLEM COMPLEXITY AND METHOD EFFICIENCY IN OPTIMIZATION
(Translated by E. R. Dawson)

Palmer GRAPHICAL EVOLUTION: AN INTRODUCTION TO THE THEORY OF RANDOM GRAPHS

Pless INTRODUCTION TO THE THEORY OF ERROR-CORRECTING CODES, SECOND EDITION

Schrijver THEORY OF LINEAR AND INTEGER PROGRAMMING

Tomescu PROBLEMS IN COMBINATORICS AND GRAPH THEORY
(Translated by R. A. Melter)

Tucker APPLIED COMBINATORICS, SECOND EDITION

KNAPSACK PROBLEMS

Algorithms and Computer Implementations

Silvano Martello

and

Paolo Toth

DEIS, University of Bologna

JOHN WILEY & SONS

Chichester · New York · Brisbane · Toronto · Singapore

Other Wiley Editorial Offices

John Wiley & Sons, Inc., 605 Third Avenue,
New York, NY 10158-0012, USA

Jacaranda Wiley Ltd, G.P.O. Box 859, Brisbane,
Queensland 4001, Australia

John Wiley & Sons (Canada) Ltd, 22 Worcester Road,
Rexdale, Ontario M9W 1L1, Canada

John Wiley & Sons (SEA) Pte Ltd, 37 Jalan Pemimpin #05-04,
Block B, Union Industrial Building, Singapore 2057

Library of Congress Cataloging-in-Publication Data:
Martello, Silvano.
Knapsack problems : algorithms and computer implementations
Silvano Martello, Paolo Toth.
p. cm. — (Wiley-Interscience series in discrete mathematics
and optimization)
Includes bibliographical references.
ISBN 0 471 92420 2
1. Computational complexity. 2. Mathematical optimization.
3. Algorithms. 4. Linear programming. 5. Integer programming.
I. Toth, Paolo. II. Title. III. Series.
QA 267.7.M37 1990
511′.8—dc20 90-12279
CIP

British Library Cataloguing in Publication Data:
Martello, Silvano
Knapsack problems : algorithms and computer
implementations.
1. Linear programming. Computation
I. Title II. Toth, Paolo
519.72

ISBN 0 471 92420 2

Printed in Great Britain by Biddles Ltd, Guildford

Contents

Preface

The development of computational complexity theory has led, in the last fifteen years, to a fascinating insight into the inherent difficulty of combinatorial optimization problems, but has also produced an undesirable side effect which can be summarized by the "equation"

$$NP\text{-}hardness = intractability,$$

thereby diminishing attention to the study of exact algorithms for NP-hard problems. However, recent results on the solution of very large instances of integer linear programming problems with special structure on the one hand, and forty years of successful use of the simplex algorithm on the other, indicate the concrete possibility of solving problems exactly through worst-case exponential-time algorithms.

This book presents a state-of-art on exact and approximate algorithms for a number of important NP-hard problems in the field of integer linear programming, which we group under the term *knapsack*. The choice of the problems reflects our personal involvement in the field, through a series of investigations over the past ten years. Hence the reader will find not only the "classical" knapsack problems (*binary, bounded, unbounded, binary multiple*), but also less familiar problems (*subset-sum, change-making*) or well-known problems which are not usually classified in the knapsack area (*generalized assignment, bin-packing*). He will find no mention, instead, of other knapsack problems (fractional, multidimensional, non-linear), and only a limited treatment of the case of generalized upper bound constraints.

The goal of the book is to fully develop an algorithmic approach without losing mathematical rigour. For each problem, we start by giving a mathematical model, discussing its relaxations and deriving procedures for the computation of bounds. We then develop approximate algorithms, approximation schemes, dynamic programming techniques and branch-and-bound algorithms. We analyse the computational complexity and the worst-case performance of bounds and approximate methods. The average performance of the computer implementations of exact and approximate algorithms is finally examined through extensive computational experiments. The Fortran codes implementing the most effective methods are provided in the included diskette. The codes are portable on virtually any computer, extensively commented and—hopefully—easy to use.

For these reasons, the book should be appreciated both by academic researchers and industrial practitioners. It should also be suitable for use as a supplementary text in courses emphasizing the theory and practice of algorithms, at the graduate or advanced undergraduate level. The exposition is in fact self-contained and is designed to introduce the reader to a methodology for developing the link between mathematical formulation and effective solution of a combinatorial optimization problem. The simpler algorithms introduced in the first chapters are in general extensively described, with numerous details on the techniques and data structures used, while the more complex algorithms of the following chapters are presented at a higher level, emphasizing the general philosophy of the different approaches. Many numerical examples are used to clarify the methodologies introduced. For the sake of clarity, all the algorithms are presented in the form of pseudo-Pascal procedures. We adopted a structured exposition for the polynomial and pseudo-polynomial procedures, but allowed a limited use of "go to" statements for the branch-and-bound algorithms. (Although this could, of course, have been avoided, the resulting exposition would, in our opinion, have been much less readable.)

We are indebted to many people who have helped us in preparing this book. Jan Karel Lenstra suggested the subject, and provided guidance and encouragement during the two years of preparation. Mauro Dell'Amico, Laureano Escudero and Matteo Fischetti read earlier versions of the manuscript, providing valuable suggestions and pointing out several errors. (We obviously retain the sole responsibility for the surviving errors.) Constructive comments were also made by Egon Balas, Martin Dyer, Ronald Graham, Peter Hammer, Ben Lageweg, Gilbert Laporte, Manfred Padberg, David Shmoys, Carlo Vercellis and Laurence Wolsey. The computational experiments and computer typesetting with the TEX system were carried out by our students Andrea Bianchini, Giovanna Favero, Marco Girardini, Stefano Gotra, Nicola Moretti, Paolo Pinetti and Mario Zacchei.

We acknowledge the financial support of the Ministero della Pubblica Istruzione and the Consiglio Nazionale delle Ricerche. Special thanks are due to the Computing Centre of the Faculty of Engineering of the University of Bologna and its Director, Roberto Guidorzi, for the facilities provided in the computational testing of the codes.

Bologna, Italy SILVANO MARTELLO
July 1989 PAOLO TOTH

1

Introduction

1.1 WHAT ARE KNAPSACK PROBLEMS?

Suppose a hitch-hiker has to fill up his knapsack by selecting from among various possible objects those which will give him maximum comfort. This *knapsack problem* can be mathematically formulated by numbering the objects from 1 to n and introducing a vector of binary variables x_j $(j = 1, \ldots, n)$ having the following meaning:

$$x_j = \begin{cases} 1 & \text{if object } j \text{ is selected;} \\ 0 & \text{otherwise.} \end{cases}$$

Then, if p_j is a measure of the comfort given by object j, w_j its size and c the size of the knapsack, our problem will be to select, from among all binary vectors x satisfying the *constraint*

$$\sum_{j=1}^{n} w_j x_j \leq c,$$

the one which *maximizes* the *objective function*

$$\sum_{j=1}^{n} p_j x_j .$$

If the reader of this book does not, or no longer practises hitch-hiking, he might be more interested in the following problem. Suppose you want to invest—all or in part—a capital of c dollars and you are considering n possible investments. Let p_j be the profit you expect from investment j, and w_j the amount of dollars it requires. It is self-evident that the optimal solution of the knapsack problem above will indicate the best possible choice of investments.

At this point you may be stimulated to solve the problem. A naive approach would be to program a computer to examine all possible binary vectors x, selecting the best of those which satisfy the constraint. Unfortunately, the number of such vectors is 2^n, so even a hypothetical computer, capable of examining one billion vectors per second, would require more than 30 years for $n = 60$, more than 60 years for $n = 61$, ten centuries for $n = 65$, and so on. However, specialized algorithms can, in most cases, solve a problem with $n = 100\,000$ in a few seconds on a mini-computer.

The problem considered so far is representative of a variety of knapsack-type problems in which a set of entities are given, each having an associated value and size, and it is desired to select one or more disjoint subsets so that the sum of the sizes in each subset does not exceed (or equals) a given bound and the sum of the selected values is maximized.

Knapsack problems have been intensively studied, especially in the last decade, attracting both theorists and practicians. The theoretical interest arises mainly from their simple structure which, on the one hand allows exploitation of a number of combinatorial properties and, on the other, more complex optimization problems to be solved through a series of knapsack-type subproblems. From the practical point of view, these problems can model many industrial situations: capital budgeting, cargo loading, cutting stock, to mention the most classical applications. In the following chapters we shall examine the most important knapsack problems, analysing relaxations and upper bounds, describing exact and approximate algorithms and evaluating their efficiency both theoretically and through computational experiments. The Fortran codes of the principal algorithms are provided in the floppy disk accompanying the book.

1.2 TERMINOLOGY

The objects considered in the previous section will generally be called *items* and their number be indicated by n. The value and size associated with the jth item will be called *profit* and *weight*, respectively, and denoted by p_j and w_j $(j = 1, \dots, n)$.

The problems considered in Chapters 2 to 5 are *single knapsack problems*, where one *container* (or *knapsack*) must be filled with an optimal subset of items. The *capacity* of such a container will be denoted by c. Chapters 6 to 8 deal with *multiple knapsack problems*, in which more than one container is available.

It is always assumed, as is usual in the literature, that profits, weights and capacities are positive integers. The results obtained, however, can easily be extended to the case of real values and, in the majority of cases, to that of nonpositive values.

The prototype problem of the previous section,

$$\text{maximize} \quad \sum_{j=1}^{n} p_j x_j$$

$$\text{subject to} \quad \sum_{j=1}^{n} w_j x_j \leq c,$$

$$x_j = 0 \text{ or } 1, \qquad j = 1, \dots, n,$$

is known as the *0-1 Knapsack Problem* and will be analysed in Chapter 2. In Section 2.12 we consider the generalization arising when the item set is partitioned into subsets and the additional constraint is imposed that at most one item per subset is selected (*Multiple-Choice Knapsack Problem*).

The problem can be generalized by assuming that for each j ($j = 1, \dots, n$), b_j items of profit p_j and weight w_j are available ($b_j \leq c/w_j$): thus we obtain the *Bounded Knapsack Problem*, defined by

$$\text{maximize} \quad \sum_{j=1}^{n} p_j x_j$$

$$\text{subject to} \quad \sum_{j=1}^{n} w_j x_j \leq c,$$

$$0 \leq x_j \leq b_j, \qquad j = 1, \dots, n,$$

$$x_j \text{ integer}, \qquad j = 1, \dots, n.$$

The problem is considered in Chapter 3. The special case in which $b_j = +\infty$ for all j (*Unbounded Knapsack Problem*) is treated in Section 3.6.

In Chapter 4 we examine the particular case of the 0-1 knapsack problem arising when $p_j = w_j$ ($j = 1, \dots, n$), as frequently occurs in practical applications. The problem is to find a subset of weights whose sum is closest to, without exceeding, the capacity, i.e.

$$\text{maximize} \quad \sum_{j=1}^{n} w_j x_j$$

$$\text{subject to} \quad \sum_{j=1}^{n} w_j x_j \leq c,$$

$$x_j = 0 \text{ or } 1, \qquad j = 1, \dots, n,$$

generally referred to as the *Subset-Sum Problem*.

In Chapter 5 a very particular bounded knapsack problem is considered, arising when $p_j = 1$ ($j = 1, \dots, n$) and, in the capacity constraint, we impose equality instead of inequality. This gives

$$\text{maximize} \quad \sum_{j=1}^{n} x_j$$

$$\text{subject to} \quad \sum_{j=1}^{n} w_j x_j = c,$$

$$0 \leq x_j \leq b_j \qquad j = 1, \dots, n,$$

$$x_j \text{ integer} \qquad j = 1, \dots, n,$$

usually called the *Change-Making Problem*, since it recalls the situation of a cashier having to assemble a given change c using the maximum (or minimum) number of coins. The same chapter deeply analyses the *Unbounded Change-Making Problem*, in which $b_j = +\infty$ for all j.

An important generalization of the 0-1 knapsack problem, discussed in Chapter 6, is the *0-1 Multiple Knapsack Problem*, arising when m containers, of given capacities c_i $(i = 1, \dots, m)$ are available. By introducing binary variables x_{ij}, taking value 1 if item j is selected for container i, 0 otherwise, we obtain the formulation

$$\text{maximize} \quad \sum_{i=1}^{m} \sum_{j=1}^{n} p_j x_{ij}$$

$$\text{subject to} \quad \sum_{j=1}^{n} w_j x_{ij} \le c_i, \qquad i = 1, \dots, m,$$

$$\sum_{i=1}^{m} x_{ij} \le 1, \qquad j = 1, \dots, n,$$

$$x_{ij} = 0 \text{ or } 1, \qquad i = 1, \dots, m, \; j = 1, \dots, n.$$

Now consider a 0-1 multiple knapsack problem in which the profit and weight of each item vary according to the container for which they are selected. By defining p_{ij} (resp. w_{ij}) as the profit (resp. the weight) of item j if inserted in container i, we get

$$\text{maximize} \quad \sum_{i=1}^{m} \sum_{j=1}^{n} p_{ij} x_{ij}$$

$$\text{subject to} \quad \sum_{j=1}^{n} w_{ij} x_{ij} \le c_i, \qquad i = 1, \dots, m,$$

$$\sum_{i=1}^{m} x_{ij} \le 1, \qquad j = 1, \dots, n,$$

$$x_{ij} = 0 \text{ or } 1, \qquad i = 1, \dots, m, \; j = 1, \dots, n,$$

known as the *Generalized Assignment Problem*, which is dealt with in Chapter 7. This is not, strictly speaking, a knapsack problem, but is included in this review because knapsack subproblems play a central role in the algorithms for solving it.

The problem is generally viewed as that of optimally assigning, all or in part, n jobs to m machines (n tasks to m agents, and so on), given the profit, p_{ij}, obtainable if machine i is assigned job j, the corresponding resource, w_{ij}, required, and the amount, c_i, of resource available to machine i.

In Chapter 8 we consider the well-known *Bin-Packing Problem*, which is not usually included in the knapsack area, but can be interpreted as a multiple subset-sum problem where all containers have the same capacity c, all items must be selected and it is desired to minimize the number of containers used. Given any upper bound m on the number of containers, and introducing m binary variables y_i, taking value 0 if container i is used, value 1 otherwise, we can state the problem as:

$$\text{maximize} \quad \sum_{i=1}^{m} y_i$$

$$\text{subject to} \quad \sum_{j=1}^{n} w_j x_{ij} \le c(1 - y_i), \qquad i = 1, \dots, m,$$

$$\sum_{i=1}^{m} x_{ij} = 1, \qquad j = 1, \dots, n,$$

$$y_i = 0 \text{ or } 1, \qquad i = 1, \dots, m,$$

$$x_{ij} = 0 \text{ or } 1, \qquad i = 1, \dots, m, j = 1, \dots, n.$$

In the last decades, an impressive amount of research on knapsack problems has been published in the literature. Reviews have been presented in the following surveys:

Salkin and De Kluyver (1975) present a number of industrial applications and results in transforming integer linear programs to knapsack problems (an approach which appeared very promising at that time);

Martello and Toth (1979) consider exact algorithms for the zero-one knapsack problem and their average computational performance; the study is extended to the other linear knapsack problems and to approximate algorithms in Martello and Toth (1987);

Dudzinski and Walukiewicz (1987) analyse dual methods for solving Lagrangian and linear programming relaxations.

In addition, almost all books on integer programming contain a section on knapsack problems. Mention is made of those by Hu (1969), Garfinkel and Nemhauser (1972), Salkin (1975), Taha (1975), Papadimitriou and Steiglitz (1982), Syslo, Deo and Kowalik (1983), Schrijver (1986), Nemhauser and Wolsey (1988).

1.3 COMPUTATIONAL COMPLEXITY

We have so far introduced the following problems:

0-1 KNAPSACK;
BOUNDED KNAPSACK;
SUBSET-SUM;
CHANGE-MAKING;
0-1 MULTIPLE KNAPSACK;
GENERALIZED ASSIGNMENT;
BIN-PACKING.

We will now show that all these problems are *NP-hard* (we refer the reader to Garey and Johnson (1979) for a thorough discussion on this concept). For each problem P, we either prove that its *recognition version* R(P) is NP-complete or that it is a generalization of a problem already proved to be NP-hard.

The following recognition problem:

PARTITION: given n positive integers $w_1, \dots, w_n$, is there a subset $S \subseteq N = \{1, \dots, n\}$ such that $\sum_{j \in S} w_j = \sum_{j \in N \setminus S} w_j$?

is a basic NP-complete problem, originally treated in Karp (1972).

(a) *SUBSET-SUM is NP-hard.*

Proof. Consider R(SUBSET-SUM), i.e.: given $n+2$ positive integers $w_1, \dots, w_n, c$ and a, is there a subset $S \subseteq N = \{1, \dots, n\}$ such that $\sum_{j \in S} w_j \leq c$ and $\sum_{j \in S} w_j \geq a$?

Any instance I of PARTITION can be *polynomially transformed* into an equivalent instance I' of R(SUBSET-SUM) by setting $c = a = \sum_{j \in N} w_j / 2$ (the answer for I is "yes" if and only if the answer for I' is "yes"). □

(b) *0-1 KNAPSACK is NP-hard.*

Proof. SUBSET-SUM is the particular case of 0-1 KNAPSACK when $p_j = w_j$ for all $j \in N$. □

(c) *BOUNDED KNAPSACK is NP-hard.*

Proof. 0-1 KNAPSACK is the particular case of BOUNDED KNAPSACK when $b_j = 1$ for all $j \in N$. □

(d) *CHANGE-MAKING is NP-hard.*

Proof. We prove NP-hardness of the special case in which $b_j = 1$ for all j. Consider R(CHANGE-MAKING), i.e.: given $n + 2$ positive integers $w_1, \dots, w_n$, c and a, is there a subset $S \subseteq N = \{1, \dots, n\}$ such that $\sum_{j \in S} w_j = c$ and $|S| \geq a$? Any instance I of PARTITION can be polynomially transformed into an equivalent instance I' of R(CHANGE-MAKING) by setting $c = \sum_{j \in N} w_j / 2$ and $a = 1$. □

Consequently, these single knapsack problems cannot be solved in a time bounded by a polynomial in n, unless $\mathcal{P} = \mathcal{NP}$. All of them, however, admit a *pseudo-polynomial* algorithm, i.e. an algorithm whose time (and space) complexity is bounded by a polynomial in n and c. In fact, it can easily be verified that the following dynamic programming recursions solve the corresponding problems. (More detailed descriptions can be found in the specific chapters.) Given any instance of a single knapsack problem, consider the sub-instance defined by items $1, \dots, j$ and capacity u ($j \leq n$, $u \leq c$). Let $f_j(u)$ be the corresponding optimal solution value ($f_j(u) = -\infty$ if no feasible solution exists) and $S_j(u)$ the optimal subset of items. The optimal solution value of the problem, $f_n(c)$, can then be obtained by iteratively applying the following recursive formulae:

0-1 KNAPSACK:

$$f_1(u) = \begin{cases} 0 & \text{for } u = 0, \dots, w_1 - 1; \\ p_1 & \text{for } u = w_1, \dots, c; \end{cases}$$

$$f_j(u) = \max(f_{j-1}(u), f_{j-1}(u - w_j) + p_j) \text{ for } j = 2, \dots, n \text{ and } u = 0, \dots, c;$$

time complexity $O(nc)$.

BOUNDED KNAPSACK:

$$f_1(u) = \begin{cases} lp_1 & \text{for } l = 0, \dots, b_1 - 1 \text{ and } u = lw_1, \dots, (l+1)w_1 - 1; \\ b_1 p_1 & \text{for } u = b_1 w_1, \dots, c; \end{cases}$$

$$f_j(u) = \max\{ f_{j-1}(u - lw_j) + lp_j : 0 \leq l \leq b_j \} \text{ for } j = 2, \dots, n \text{ and } u = 0, \dots, c;$$

time complexity $O(c \sum_{j=1}^{n} b_j)$, that is, in the worst case, $O(nc^2)$.

SUBSET-SUM:

Same as 0-1 KNAPSACK, but with p_j replaced by w_j.

CHANGE-MAKING:

$$f_1(u) = \begin{cases} l & \text{for } u = lw_1, \text{ with } l = 0, \ldots, b_1; \\ -\infty & \text{for all positive } u \leq c \text{ such that } u (\text{mod } w_1) \neq 0 \,; \end{cases}$$

$$f_j(u) = \max\{ f_{j-1}(u - lw_j) + l : \ 0 \leq l \leq b_j \} \text{ for } j = 2, \ldots, n \text{ and } u = 0, \ldots, c;$$

time complexity $O(c \sum_{j=1}^{n} b_j)$, that is, in the worst case, $O(nc^2)$.

For all the algorithms the computation of $S_j(u)$ is straightforward. Since, for each j, we only need to store $S_{j-1}(u)$ and $S_j(u)$ for all u, the space complexity is always $O(nc)$.

For the multiple problems (0-1 MULTIPLE KNAPSACK, GENERALIZED ASSIGNMENT, BIN-PACKING) no pseudo-polynomial algorithm can exist, unless $\mathcal{P} = \mathcal{NP}$, since the problems can be proved to be *NP-hard in the strong sense*. Consider in fact the following recognition problem:

3-PARTITION: given $n = 3m$ positive integers $w_1, \ldots, w_n$ satisfying $\sum_{j=1}^{n} w_j/m = B$ integer and $B/4 < w_j < B/2$ for $j = 1, \ldots, n$, is there a partition of $N = \{1, \ldots, n\}$ into m subsets $S_1, \ldots, S_m$ such that $\sum_{j \in S_i} w_j = B$ for $i = 1, \ldots, m$? (Notice that each S_i must contain exactly three elements from N.)

This is the first problem discovered to be NP-complete in the strong sense (Garey and Johnson, 1975).

(e) *0-1 MULTIPLE KNAPSACK is NP-hard in the strong sense.*

Proof. Consider R(0-1 MULTIPLE KNAPSACK), i.e.: given $2n + m + 1$ positive integers: $p_1, \ldots, p_n$; $w_1, \ldots, w_n$; $c_1, \ldots, c_m$, and a, are there m disjoint subsets $S_1, \ldots, S_m$ of $N = \{1, \ldots, n\}$ such that $\sum_{j \in S_i} w_j \leq c_i$ for $i = 1, \ldots, m$ and $\sum_{i=1}^{m} \sum_{j \in S_i} p_j \geq a$? Any instance I of 3-PARTITION can be *pseudo-polynomially transformed* into an equivalent instance I' of R(0-1 MULTIPLE KNAPSACK) by setting $c_i = B$ for $i = 1, \ldots, m$, $p_j = 1$ for $j = 1, \ldots, n$ and $a = n$ (which implies that $\bigcup_{i=1}^{m} S_i = N$ in any "yes" instance). □

(f) *GENERALIZED ASSIGNMENT is NP-hard in the strong sense.*

Proof. Immediate, since 0-1 MULTIPLE KNAPSACK is the particular case of GENERALIZED ASSIGNMENT when $p_{ij} = p_j$ and $w_{ij} = w_j$ for $i = 1, \ldots, m$ and $j = 1, \ldots, n$. □

(g) *BIN-PACKING is NP-hard in the strong sense.*

Proof. Consider R(BIN-PACKING), i.e.: given $n+2$ positive integers $w_1, \dots, w_n$, c and a, is there a partition of $N = \{1, \dots, n\}$ into a subsets $S_1, \dots, S_a$ such that $\sum_{j \in S_i} w_j \leq c$ for $i = 1, \dots, a$? Any instance I of 3-PARTITION can be pseudo-polynomially transformed into an equivalent instance I' of R(BIN-PACKING) by setting $c = B$ and $a = m$. □

1.4 LOWER AND UPPER BOUNDS

In the previous section we have proved that none of our problems can be solved in polynomial time, unless $\mathcal{P} = \mathcal{NP}$. Hence in the following chapters we analyse:

(a) enumerative algorithms (having, in the worst case, running times which grow exponentially with the input size) to determine optimal solutions;

(b) approximate algorithms (with running times bounded by a polynomial in the input size) to determine feasible solutions whose value is a *lower bound* on the optimal solution value.

The average running times of such algorithms are experimentally evaluated through execution of the corresponding computer codes on different classes of randomly-generated test problems. It will be seen that the average behaviour of the enumerative algorithms is in many cases much better than the worst-case bound, allowing optimal solution of large-size problems with acceptable running times.

The *performance* of an approximate algorithm for a specific instance is measured through the ratio between the solution value found by the algorithm and the optimal solution value (notice that, for a maximization problem, this ratio is no greater than one). Besides the experimental evaluation, it is useful to provide, when possible, a theoretical measure of performance through *worst-case* analysis (see Fisher (1980) for a general introduction to this concept).

Let A be an approximate algorithm for a given maximization problem (all our considerations extend easily to the minimization case). For any instance I of the problem, let $OPT(I)$ be the optimal solution value and $A(I)$ the value found by A. Then, the *worst-case performance ratio* of A is defined as the largest real number $r(A)$ such that

$$\frac{A(I)}{OPT(I)} \geq r(A) \qquad \text{for all instances } I;$$

the closer $r(A)$ is to one, the better the worst-case behaviour of A. The proof that a given value r is the worst-case performance ratio of an algorithm A consists, in general, of two phases:

(i) it is first proved that, for any instance I of the problem, inequality $A(I)/OPT(I) \geq r$ holds;

(ii) in order to ensure that r is the largest value satisfying the inequality, i.e. that r is *tight*, a specific instance I' is produced for which $A(I')/OPT(I') = r$ holds (or a series of instances for which the above ratio tends to be arbitrarily close to r).

The performance of A can be equivalently expressed in terms of *worst-case relative error*, i.e. the smallest real number $\varepsilon(A)$ such that

$$\frac{OPT(I) - A(I)}{OPT(I)} \leq \varepsilon(A) \qquad \text{for all instances } I,$$

(i.e. $r(A) = 1 - \varepsilon(A)$).

An *approximation scheme* for a maximization problem is an algorithm A which, given an instance I and an error bound $\varepsilon > 0$, returns a solution of value $A(I)$ such that $(OPT(I) - A(I))/OPT(I) \leq \varepsilon$. Let *length* (I) denote the *input size*, i.e. the number of symbols required for coding I. If, for any fixed ε, the running time of A is bounded by a polynomial in *length* (I), then A is a *polynomial-time approximation scheme*: any relative error can be obtained in a time which is polynomial in *length* (I) (but can be exponential in $1/\varepsilon$). If the running time of A is polynomial both in *length* (I) and $1/\varepsilon$, then A is a *fully polynomial-time approximation scheme*.

In subsequent chapters we describe the most interesting polynomial-time and fully polynomial-time approximation schemes for single knapsack problems. For the remaining (multiple) problems, no fully polynomial-time approximation scheme can exist, unless $\mathcal{P} = \mathcal{NP}$, since (see Garey and Johnson (1975)) this would imply the existence of a pseudo-polynomial algorithm for their optimal solution (which is impossible, these being NP-hard problems in the strong sense). For BIN-PACKING, also the existence of a polynomial-time approximation scheme can be ruled out, unless $\mathcal{P} = \mathcal{NP}$ (Johnson, Demers, Ullman, Garey and Graham, 1974). The same holds for GENERALIZED ASSIGNMENT and 0-1 MULTIPLE KNAPSACK in the minimization version (Sahni and Gonzalez, 1976). For the maximization version of these two problems no polynomial-time approximation scheme is known, although there is no proof that it cannot exist (the proof in Sahni and Gonzalez (1976) does not extend to the maximization case).

Besides experimental and worst-case analysis, an approximate algorithm can allow *probabilistic analysis*. Speaking informally this consists of specifying an average problem instance in terms of a probability distribution over the class of all instances and evaluating running time and solution value as random variables. Examples of this approach which, however, is generally possible only for very simple algorithms, are given in Sections 2.8.3 and 4.3.4 (see Karp, Lenstra, McDiarmid and Rinnooy Kan (1985) and Rinnooy Kan (1987) for a general introduction to probabilistic analysis).

For a maximization problem, the solution value determined by an approximate algorithm limits the optimal solution value from below. It is always convenient to have methods for limiting this value from above, too. *Upper bounds* are extremely useful

(a) in enumerative algorithms, to exclude computations which cannot lead to the optimal solution;

(b) in approximate algorithms, to "a-posteriori" evaluate the performance obtained. Suppose algorithm A is applied to instance I, and let $U(I)$ be any upper bound on $OPT(I)$: it is then clear that the relative error of the approximate solution is no greater than $(U(I) - A(I))/U(I)$.

The worst-case performance ratio of an upper bounding procedure U can be defined similarly to that of an approximate algorithm, i.e. as the smallest real number $\rho(U)$ such that

$$\frac{U(I)}{OPT(I)} \leq \rho(U) \qquad \text{for all instances } I.$$

The closer $\rho(U)$ is to one, the better the worst-case behaviour of U.

Upper bounds are usually computed by solving *relaxations* of the given problems. *Continuous, Lagrangian* and *surrogate* relaxations are the most frequently used. For a given problem P, the corresponding relaxed problem will be denoted with $C(P)$, $L(P, m)$ and $S(P, m)$, m being an appropriate vector of *multipliers*. The optimal solution value of problem P will be denoted with $z(P)$.

2

0-1 Knapsack problem

2.1 INTRODUCTION

The *0-1*, or *Binary*, *Knapsack Problem* (KP) is: given a set of n *items* and a *knapsack*, with

$$p_j = \textit{profit} \text{ of item } j,$$

$$w_j = \textit{weight} \text{ of item } j,$$

$$c = \textit{capacity} \text{ of the knapsack},$$

select a subset of the items so as to

$$\text{maximize} \quad z = \sum_{j=1}^{n} p_j x_j \tag{2.1}$$

$$\text{subject to} \quad \sum_{j=1}^{n} w_j x_j \leq c, \tag{2.2}$$

$$x_j = 0 \text{ or } 1, \qquad j \in N = \{1, \ldots, n\}, \tag{2.3}$$

where

$$x_j = \begin{cases} 1 & \text{if item } j \text{ is selected;} \\ 0 & \text{otherwise.} \end{cases}$$

KP is the most important knapsack problem and one of the most intensively studied discrete programming problems. The reason for such interest basically derives from three facts: (a) it can be viewed as the simplest *Integer Linear Programming* problem; (b) it appears as a subproblem in many more complex problems; (c) it may represent a great many practical situations. Recently, it has been used for generating minimal cover induced constraints (see, e.g., Crowder, Johnson and Padberg, (1983)) and in several coefficient reduction procedures for strengthening LP bounds in general integer programming (see, e.g., Dietrich and Escudero, (1989a, 1989b)). During the last few decades, KP has been studied through different approaches, according to the theoretical development of *Combinatorial Optimization*.

In the fifties, Bellman's *dynamic programming* theory produced the first algorithms to exactly solve the 0-1 knapsack problem. In 1957 Dantzig gave an elegant and efficient method to determine the solution to the continuous relaxation of the problem, and hence an *upper bound* on z which was used in the following twenty years in almost all studies on KP.

In the sixties, the dynamic programming approach to the KP and other knapsack-type problems was deeply investigated by Gilmore and Gomory. In 1967 Kolesar experimented with the first *branch-and-bound* algorithm for the problem.

In the seventies, the branch-and-bound approach was further developed, proving to be the only method capable of solving problems with a high number of variables. The most well-known algorithm of this period is due to Horowitz and Sahni. In 1973 Ingargiola and Korsh presented the first *reduction procedure*, a preprocessing algorithm which significantly reduces the number of variables. In 1974 Johnson gave the first *polynomial-time approximation scheme* for the subset-sum problem; the result was extended by Sahni to the 0-1 knapsack problem. The first *fully polynomial-time approximation scheme* was obtained by Ibarra and Kim in 1975. In 1977 Martello and Toth proposed the first upper bound dominating the value of the continuous relaxation.

The main results of the eighties concern the solution of large-size problems, for which sorting of the variables (required by all the most effective algorithms) takes a very high percentage of the running time. In 1980 Balas and Zemel presented a new approach to solve the problem by sorting, in many cases, only a small subset of the variables (the *core problem*).

In this chapter we describe the main results outlined above in logical (not necessarily chronological) sequence. Upper bounds are described in Sections 2.2 and 2.3, approximate algorithms in Sections 2.4 and 2.8, exact algorithms in Sections 2.5, 2.6 and 2.9, reduction procedures in Section 2.7. Computational experiments are reported in Section 2.10, while Section 2.11 contains an introduction to the facetial analysis of the problem. Section 2.12 deals with a generalization of KP (the multiple-choice knapsack problem).

We will assume, without any loss of generality, that

$$p_j, w_j \text{ and } c \text{ are positive integers,} \tag{2.4}$$

$$\sum_{j=1}^{n} w_j > c, \tag{2.5}$$

$$w_j \leq c \text{ for } j \in N. \tag{2.6}$$

If assumption (2.4) is violated, fractions can be handled by multiplying through by a proper factor, while nonpositive values can be handled as follows (Glover, 1965):

1. for each $j \in N^0 = \{j \in N : p_j \leq 0, w_j \geq 0\}$ **do** $x_j := 0$;
2. for each $j \in N^1 = \{j \in N : p_j \geq 0, w_j \leq 0\}$ **do** $x_j := 1$;

3. let $N^- = \{j \in N : p_j < 0,\ w_j < 0\}$, $N^+ = N \backslash (N^0 \cup N^1 \cup N^-)$, and

$$\begin{cases} y_j = 1 - x_j,\ \overline{p}_j = -p_j,\ \overline{w}_j = -w_j & \text{for } j \in N^-, \\ y_j = x_j,\ \overline{p}_j = p_j,\ \overline{w}_j = w_j & \text{for } j \in N^+; \end{cases}$$

4. solve the residual problem

$$\text{maximize} \quad z = \sum_{j \in N^- \cup N^+} \overline{p}_j y_j + \sum_{j \in N^1 \cup N^-} p_j$$

$$\text{subject to} \quad \sum_{j \in N^- \cup N^+} \overline{w}_j y_j \le c - \sum_{j \in N^1 \cup N^-} w_j,$$

$$y_j = 0 \text{ or } 1, \qquad j \in N^- \cup N^+.$$

If the input data violate assumption (2.5) then, trivially, $x_j = 1$ for all $j \in N$; if they violate assumption (2.6), then $x_j = 0$ for each j such that $w_j > c$.

Unless otherwise specified, we will always suppose that the items are ordered according to decreasing values of the profit per unit weight, i.e. so that

$$\frac{p_1}{w_1} \ge \frac{p_2}{w_2} \ge \cdots \ge \frac{p_n}{w_n}. \tag{2.7}$$

If this is not the case, profits and weights can be re-indexed in $O(n \log n)$ time through any efficient sorting procedure (see, for instance, Aho, Hopcroft and Ullman, (1983)).

Given any problem instance I, we denote the value of any optimal solution with $z(I)$, or, when no confusion arises, with z.

KP is always considered here in maximization form. The minimization version of the problem,

$$\text{minimize} \quad \sum_{j=1}^{n} p_j y_j$$

$$\text{subject to} \quad \sum_{j=1}^{n} w_j y_j \ge q,$$

$$y_j = 0 \text{ or } 1, \qquad j \in N$$

can easily be transformed into an equivalent maximization form by setting $y_j = 1 - x_j$ and solving (2.1), (2.2), (2.3) with $c = \sum_{j=1}^{n} w_j - q$. Let $z\max$ be the solution value of such a problem: the minimization problem then has solution value $z\min = \sum_{j=1}^{n} p_j - z\max$. (Intuitively, we maximize the total profit of the items *not inserted* in the knapsack.)

2.2 RELAXATIONS AND UPPER BOUNDS

2.2.1 Linear programming relaxation and Dantzig's bound

The most natural, and historically the first, relaxation of KP is the *linear programming relaxation*, i.e. the *continuous knapsack problem* $C(KP)$ obtained from (2.1), (2.2), (2.3) by removing the integrality constraint on x_j:

$$\text{maximize} \quad \sum_{j=1}^{n} p_j x_j$$

$$\text{subject to} \quad \sum_{j=1}^{n} w_j x_j \leq c,$$

$$0 \leq x_j \leq 1, \qquad j = 1, \dots, n.$$

Suppose that the items, ordered according to (2.7), are consecutively inserted into the knapsack until the first item, s, is found which does not fit. We call it the *critical item*, i.e.

$$s = \min \left\{ j : \sum_{i=1}^{j} w_i > c \right\}, \tag{2.8}$$

and note that, because of assumptions (2.4)–(2.6), we have $1 < s \leq n$. Then $C(KP)$ can be solved through a property established by Dantzig (1957), which can be formally stated as follows.

Theorem 2.1 *The optimal solution* $\overline{x}$ *of* $C(KP)$ *is*

$$\overline{x}_j = 1 \qquad \text{for } j = 1, \dots, s-1,$$

$$\overline{x}_j = 0 \qquad \text{for } j = s+1, \dots, n,$$

$$\overline{x}_s = \frac{\overline{c}}{w_s},$$

where

$$\overline{c} = c - \sum_{j=1}^{s-1} w_j. \tag{2.9}$$

Proof. A graphical proof can be found in Dantzig (1957). More formally, observe that any optimal solution x of $C(KP)$ must be maximal, in the sense that $\sum_{j=1}^{n} w_j x_j = c$. Assume, without loss of generality, that $p_j / w_j > p_{j+1} / w_{j+1}$ for all j, and let x^* be the optimal solution of $C(KP)$. Suppose, by absurdity, that $x_k^* < 1$ for some $k < s$, then we must have $x_q^* > \overline{x}_q$ for at least one item $q \geq s$.

Given a sufficiently small $\varepsilon > 0$, we could increase the value of x_k^* by ε and decrease that of x_q^* by $\varepsilon w_k/w_q$, thus augmenting the value of the objective function of $\varepsilon(p_k - p_q w_k/w_q)$ (> 0, since $p_k/w_k > p_q/w_q$), which is a contradiction. In the same way we can prove that $x_k^* > 0$ for $k > s$ is impossible. Hence $\bar{x}_s = \bar{c}/w_s$ from maximality. □

The optimal solution value of $C(KP)$ follows:

$$z(C(KP)) = \sum_{j=1}^{s-1} p_j + \bar{c}\frac{p_s}{w_s}.$$

Because of the integrality of p_j and x_j, a valid upper bound on $z(KP)$ is thus

$$U_1 = \lfloor z(C(KP)) \rfloor = \sum_{j=1}^{s-1} p_j + \left\lfloor \bar{c}\frac{p_s}{w_s} \right\rfloor, \tag{2.10}$$

where $\lfloor a \rfloor$ denotes the largest integer not greater than a.

The worst-case performance ratio of U_1 is $\rho(U_1) = 2$. This can easily be proved by observing that, from (2.10), $U_1 < \sum_{j=1}^{s-1} p_j + p_s$. Both $\sum_{j=1}^{s-1} p_j$ and p_s are feasible solution values for KP, hence no greater than the optimal solution value z, thus, for any instance, $U_1/z < 2$. To see that $\rho(U_1)$ is tight, consider the series of problems with $n = 2$, $p_1 = w_1 = p_2 = w_2 = k$ and $c = 2k - 1$, for which $U_1 = 2k - 1$ and $z = k$, so U_1/z can be arbitrarily close to 2 for k sufficiently large.

The computation of $z(C(KP))$, hence that of the Dantzig bound U_1, clearly requires $O(n)$ time if the items are already sorted as assumed. If this is not the case, the computation can still be performed in $O(n)$ time by using the following procedure to determine the critical item.

2.2.2 Finding the critical item in $O(n)$ time

For each $j \in N$, define $r_j = p_j/w_j$. The *critical ratio* r_s can then be identified by determining a partition of N into $J1 \cup JC \cup J0$ such that

$$\begin{aligned} r_j &> r_s && \text{for } j \in J1, \\ r_j &= r_s && \text{for } j \in JC, \\ r_j &< r_s && \text{for } j \in J0, \end{aligned}$$

and

$$\sum_{j \in J1} w_j \le c < \sum_{j \in J1 \cup JC} w_j.$$

The procedure, proposed by Balas and Zemel (1980), progressively determines $J1$

and $J0$ using, at each iteration, a tentative value λ to partition the set of currently "free" items $N \backslash (J1 \cup J0)$. Once the final partition is known, the critical item s is identified by filling the residual capacity $c - \sum_{j \in J1} w_j$ with items in JC, in any order.

```
procedure CRITICAL_ ITEM:
input: n, c, (p_j), (w_j);
output: s;
begin
    J1 := Ø;
    J0 := Ø;
    JC := {1, ..., n};
    c̄ := c;
    partition := "no";
    while partition = "no" do
        begin
            determine the median λ of the values in R = {p_j/w_j : j ∈ JC};
            G := {j ∈ JC : p_j/w_j > λ};
            L := {j ∈ JC : p_j/w_j < λ};
            E := {j ∈ JC : p_j/w_j = λ};
            c' := Σ_{j∈G} w_j;
            c'' := c' + Σ_{j∈E} w_j;
            if c' ≤ c̄ < c'' then partition := "yes"
            else if c' > c̄ then (comment: λ is too small)
                begin
                    J0 := J0 ∪ L ∪ E;
                    JC := G
                end
            else (comment: λ is too large)
                begin
                    J1 := J1 ∪ G ∪ E;
                    JC := L;
                    c̄ := c̄ − c''
                end
        end;
    J1 := J1 ∪ G;
    J0 := J0 ∪ L;
    JC := E (= {e_1, e_2, ..., e_q});
    c̄ := c̄ − c';
    σ := min {j : Σ_{i=1}^{j} w_{e_i} > c̄};
    s := e_σ
end.
```

Finding the median of m elements requires $O(m)$ time (see Aho, Hopcroft and Ullman, (1983)), so each iteration of the "while" loop requires $O(|JC|)$ time. Since at least half the elements of JC are eliminated at each iteration, the overall time complexity of the procedure is $O(n)$.

The solution of $C(KP)$ can then be determined as

$$\overline{x}_j = 1 \qquad \text{for } j \in J1 \cup \{e_1, e_2, \ldots, e_{\sigma-1}\};$$

$$\overline{x}_j = 0 \qquad \text{for } j \in J0 \cup \{e_{\sigma+1}, \ldots, e_q\};$$

$$\overline{x}_s = \left(c - \sum_{j \in N \setminus \{s\}} w_j \overline{x}_j \right) / w_s.$$

2.2.3 Lagrangian relaxation

An alternative way to relax KP is through the Lagrangian approach. Given a non-negative multiplier λ, the *Lagrangian relaxation* of KP ($L(KP, \lambda)$) is

$$\text{maximize} \quad \sum_{j=1}^{n} p_j x_j + \lambda \left(c - \sum_{j=1}^{n} w_j x_j \right)$$

$$\text{subject to} \quad x_j = 0 \text{ or } 1, \qquad j = 1, \ldots, n.$$

The objective function can be restated as

$$z(L(KP, \lambda)) = \sum_{j=1}^{n} \tilde{p}_j x_j + \lambda c, \tag{2.11}$$

where $\tilde{p}_j = p_j - \lambda w_j$ for $j = 1, \ldots, n$, and the optimal solution of $L(KP, \lambda)$ is easily determined, in $O(n)$ time, as

$$\tilde{x}_j = \begin{cases} 1 & \text{if } \tilde{p}_j > 0, \\ 0 & \text{if } \tilde{p}_j < 0. \end{cases} \tag{2.12}$$

(When $\tilde{p}_j = 0$, the value of $\tilde{x}_j$ is immaterial.) Hence, by defining $J(\lambda) = \{j : p_j / w_j > \lambda\}$, the solution value of $L(KP, \lambda)$ is

$$z(L(KP, \lambda)) = \sum_{j \in J(\lambda)} \tilde{p}_j + \lambda c.$$

For any $\lambda \geq 0$, this is an upper bound on $z(KP)$ which, however, can never be better than the Dantzig bound U_1. In fact (2.12) also gives the solution of the continuous relaxation of $L(KP, \lambda)$, so

$$z(L(KP, \lambda)) = z(C(L(KP, \lambda))) \geq z(C(KP)).$$

The value of λ producing the minimum value of $z(L(KP, \lambda))$ is $\lambda^* = p_s/w_s$. With this value, in fact, we have $\tilde{p}_j \geq 0$ for $j = 1, \ldots, s-1$ and $\tilde{p}_j \leq 0$ for $j = s, \ldots, n$, so $J(\lambda^*) \subseteq \{1, \ldots, s-1\}$. Hence $\tilde{x}_j = \overline{x}_j$ for $j \in N \backslash \{s\}$ (where $(\overline{x}_j)$ is defined by Theorem 2.1) and, from (2.11)–(2.12), $z(L(KP, \lambda^*)) = \sum_{j=1}^{s-1}(p_j - \lambda^* w_j) + \lambda^* c = z(C(KP))$. Also notice that, for $\lambda = \lambda^*$, $\tilde{p}_j$ becomes

$$p_j^* = p_j - w_j \frac{p_s}{w_s}; \tag{2.13}$$

$| p_j^* |$ is the decrease which we obtain in $z(L(KP, \lambda^*))$ by setting $\tilde{x}_j = 1 - \tilde{x}_j$, and hence a *lower bound* on the corresponding decrease in the continuous solution value (since the optimal λ generally changes by imposing the above conditions). The value of $| p_j^* |$ will be very useful in the next sections.

Other properties of the Lagrangian relaxation for KP have been investigated by Maculan (1983). See also Fisher (1981) for a general survey on Lagrangian relaxations.

2.3 IMPROVED BOUNDS

In the present section we consider upper bounds dominating the Dantzig one, useful to improve on the average efficiency of algorithms for KP. Because of this dominance property, the worst-case performance ratio of these bounds is at most 2. Indeed, it is exactly 2, as can easily be verified through series of examples similar to that introduced for U_1, i.e. having $p_j = w_j$ for all j (so that the bounds take the trivial value c).

2.3.1 Bounds from additional constraints

Martello and Toth obtained the first upper bound dominating the Dantzig one, by imposing the integrality of the *critical variable* x_s.

Theorem 2.2 (Martello and Toth, 1977a) *Let*

$$U^0 = \sum_{j=1}^{s-1} p_j + \left\lfloor \overline{c} \frac{p_{s+1}}{w_{s+1}} \right\rfloor, \tag{2.14}$$

$$U^1 = \sum_{j=1}^{s-1} p_j + \left\lfloor p_s - (w_s - \overline{c}) \frac{p_{s-1}}{w_{s-1}} \right\rfloor, \tag{2.15}$$

where s and $\overline{c}$ are the values defined by (2.8) and (2.9). Then

(i) an upper bound on $z(KP)$ is

$$U_2 = \max(U^0, U^1); \tag{2.16}$$

(ii) for any instance of KP, we have $U_2 \le U_1$.

Proof. (i) Since x_s cannot take a fractional value, the optimal solution of KP can be obtained from the continuous solution $\overline{x}$ of $C(KP)$ either without inserting item s (i.e. by imposing $\overline{x}_s = 0$), or by inserting it (i.e. by imposing $\overline{x}_s = 1$) and hence removing at least one of the first $s - 1$ items. In the former case, the solution value cannot exceed U^0, which corresponds to the case of filling the residual capacity $\overline{c}$ with items having the best possible value of p_j/w_j (i.e. p_{s+1}/w_{s+1}). In the latter it cannot exceed U^1, where it is supposed that the item to be removed has exactly the minimum necessary value of w_j (i.e. $w_s - \overline{c}$) and the worst possible value of p_j/w_j (i.e. p_{s-1}/w_{s-1}).

(ii) $U^0 \le U_1$ directly follows from (2.10), (2.14) and (2.7). To prove that $U^1 \le U_1$ also holds, notice that $p_s/w_s \le p_{s-1}/w_{s-1}$ (from (2.7)), and $\overline{c} < w_s$ (from (2.8), (2.9)). Hence

$$(\overline{c} - w_s)\left(\frac{p_s}{w_s} - \frac{p_{s-1}}{w_{s-1}}\right) \ge 0,$$

and, by algebraic manipulation,

$$\overline{c}\frac{p_s}{w_s} \ge p_s - (w_s - \overline{c})\frac{p_{s-1}}{w_{s-1}},$$

from which one has the thesis. □

The time complexity for the computation of U_2 is trivially $O(n)$, once the critical item is known.

Example 2.1

Consider the instance of KP defined by

$n = 8,$

$(p_j) = (15, 100, 90, 60, 40, 15, 10, 1),$

$(w_j) = (2, 20, 20, 30, 40, 30, 60, 10),$

$c = 102.$

The optimal solution is $x = (1, 1, 1, 1, 0, 1, 0, 0)$, of value $z = 280$. From (2.8) we have $s = 5$. Hence

$$U_1 = 265 + \left\lfloor 30 \frac{40}{40} \right\rfloor = 295.$$

$$U^0 = 265 + \left\lfloor 30 \frac{15}{30} \right\rfloor = 280;$$

$$U^1 = 265 + \left\lfloor 40 - 10 \frac{60}{30} \right\rfloor = 285;$$

$$U_2 = 285. \square$$

The consideration on which the Martello and Toth bound is based can be further exploited to compute more restrictive upper bounds than U_2. This can be achieved by replacing the values U^0 and U^1 with tighter values, say $\overline{U}^0$ and $\overline{U}^1$, which take the exclusion and inclusion of item s more carefully into account. Hudson (1977) has proposed computing $\overline{U}^1$ as the solution value of the continuous relaxation of KP with the additional constraint $x_s = 1$. Fayard and Plateau (1982) and, independently, Villela and Bornstein (1983), proposed computing $\overline{U}^0$ as the solution value of $C(KP)$ with the additional constraint $x_s = 0$.

By defining $\sigma^1(j)$ and $\sigma^0(j)$ as the critical item when we impose, respectively, $x_j = 1$ $(j \geq s)$ and $x_j = 0$ $(j \leq s)$, that is

$$\sigma^1(j) = \min \left\{ k : \sum_{i=1}^{k} w_i > c - w_j \right\}, \tag{2.17}$$

$$\sigma^0(j) = \min \left\{ k : \sum_{\substack{i=1 \\ i \neq j}}^{k} w_i > c \right\}, \tag{2.18}$$

we obtain

$$\overline{U}^0 = \sum_{\substack{j=1 \\ j \neq s}}^{\sigma^0(s)-1} p_j + \left\lfloor \left(c - \sum_{\substack{j=1 \\ j \neq s}}^{\sigma^0(s)-1} w_j \right) \frac{p_{\sigma^0(s)}}{w_{\sigma^0(s)}} \right\rfloor; \tag{2.19}$$

$$\overline{U}^1 = p_s + \sum_{j=1}^{\sigma^1(s)-1} p_j + \left\lfloor \left(c - w_s - \sum_{j=1}^{\sigma^1(s)-1} w_j \right) \frac{p_{\sigma^1(s)}}{w_{\sigma^1(s)}} \right\rfloor; \tag{2.20}$$

and the new upper bound

$$U_3 = \max\left(\overline{U}^0, \overline{U}^1\right).$$

It is self-evident that:

(a) $\overline{U}^0 \leq U^0$ and $\overline{U}^1 \leq U^1$, so $U_3 \leq U_2$;

(b) the time complexity for the computation of U_3 is the same as for U_1 and U_2, i.e. $O(n)$.

Example 2.1 (continued)

From (2.17)–(2.20) we have

$$\sigma^0(5) = 7, \quad \overline{U}^0 = 280 + \left\lfloor 0 \frac{10}{60} \right\rfloor = 280;$$

$$\sigma^1(5) = 4, \quad \overline{U}^1 = 40 + 205 + \left\lfloor 20 \frac{60}{30} \right\rfloor = 285;$$

$$U_3 = 285. \square$$

2.3.2 Bounds from Lagrangian relaxations

Other bounds computable in $O(n)$ time can be described through the terminology introduced in Section 2.2 for the Lagrangian relaxation of the problem. Remember that $z(C(KP)) = z(L(KP, \lambda^*))$ and $|\, p_j^* \,|$ (see 2.13) is a lower bound on the decrease of $z(C(KP))$ corresponding to the change of the jth variable from $\tilde{x}_j$ to $1 - \tilde{x}_j$. Müller-Merbach (1978) noted that, in order to obtain an integer solution from the continuous one, either (a) the fractional variable $\overline{x}_s$ alone has to be reduced to 0 (without any changes of the other variables), or (b) at least one of the other variables, say $\overline{x}_j$, has to change its value (from 1 to 0 or from 0 to 1). In case (a) the value of $z(C(KP))$ decreases by $\overline{c}p_s/w_s$, in case (b) by at least $|\, p_j^* \,|$. Hence the Müller-Merbach bound

$$U_4 = \max \left(\sum_{j=1}^{s-1} p_j, \max \left\{ \lfloor z(C(KP)) - |\, p_j^* \,| \rfloor : j \in N \backslash \{s\} \right\} \right). \tag{2.21}$$

It is immediately evident that $U_4 \leq U_1$. No dominance exists, instead, between U_4 and the other bounds. For the instance of example 2.1 we have $U_3 = U_2 < U_4$ (see below), but it is not difficult to find examples (see Müller-Merbach (1978)) for which $U_4 < U_3 \leq U_2$.

The ideas behind bounds U_2, U_3 and U_4 have been further exploited by Dudzinski and Walukiewicz (1984a), who have obtained an upper bound dominating all the above. Consider any feasible solution $\hat{x}$ to KP that we can obtain from the continuous one as follows:

1. **for each** $k \in N \backslash \{s\}$ **do** $\hat{x}_k := \overline{x}_k$;
2. $\hat{x}_s := 0$;

3. **for each** k such that $\hat{x}_k = 0$ **do**
   if $w_k \leq c - \sum_{j=1}^{n} w_j \hat{x}_j$ **then** $\hat{x}_k := 1$,

and define $\hat{N} = \{j \in N \backslash \{s\} : \hat{x}_j = 0\}$ ($\hat{x}$ is closely related to the *greedy solution*, discussed in Section 2.4). Noting that an optimal integer solution can be obtained (a) by setting $x_s = 1$ or (b) by setting $x_s = 0$ and $x_j = 1$ for at least one $j \in \hat{N}$, it is immediate to obtain the Dudzinski and Walukiewicz (1984a) bound:

$$U_5 = \max\Big(\min\big(\overline{U}^{\,1}, \max\{\lfloor z(C(KP)) - p_j^* \rfloor : j = 1, \dots, s-1\}\big), \min\big(\overline{U}^{\,0}, \max\{\lfloor z(C(KP)) + p_j^* \rfloor : j \in \hat{N}\}\big), \sum_{j=1}^{n} p_j \hat{x}_j\Big), \tag{2.22}$$

where $\overline{U}^{\,0}$ and $\overline{U}^{\,1}$ are given by (2.19) and (2.20). The time complexity is $O(n)$.

Example 2.1 (continued)

From (2.13), $(p_j{}^*) = (13, 80, 70, 30, 0, -15, -50, -9)$. Hence:

$U_4 = \max(265, \max\{282, 215, 225, 265, 280, 245, 286\}) = 286.$

$(\hat{x}_j) = (1, 1, 1, 1, 0, 1, 0, 0);$

$U_5 = \max(\min(285, \max\{282, 215, 225, 265\}),$
$\min(280, \max\{245, 286\}), 280) = 282.$ □

2.3.3 Bounds from partial enumeration

Bound U_3 of Section 2.3.1 can also be seen as the result of the application of the Dantzig bound at the two terminal nodes of a decision tree having the root node corresponding to KP and two descendent nodes, say N0 and N1, corresponding to the exclusion and inclusion of item s. Clearly, the maximum among the upper bounds corresponding to all the terminal nodes of a decision tree represents a valid upper bound for the original problem corresponding to the root node. So, if $\overline{U}^{\,0}$ and $\overline{U}^{\,1}$ are the Dantzig bounds corresponding respectively to nodes N0 and N1, U_3 represents a valid upper bound for KP.

An improved bound, U_6, can be obtained by considering decision trees having more than two terminal nodes, as proposed by Martello and Toth (1988).

In order to introduce this bound, suppose s has been determined, and let r, t be any two items such that $1 < r \leq s$ and $s \leq t < n$. We can obtain a feasible solution for KP by setting $x_j = 1$ for $j < r$, $x_j = 0$ for $j > t$ and finding the optimal solution of subproblem KP(r, t) defined by items $r, r+1, \ldots, t$ with reduced capacity $c(r) = c - \sum_{j=1}^{r-1} w_j$. Suppose now that KP($r, t$) is solved through an elementary binary decision-tree which, for $j = r, r+1, \ldots, t$, generates pairs of decision nodes by setting, respectively, $x_j = 1$ and $x_j = 0$; each node k (obtained, say, by fixing x_j) generates a pair of descendent nodes (by fixing x_{j+1}) iff $j < t$ and the solution corresponding to k is feasible. For each node k of the resulting tree, let $f(k)$ be the item from which k has been generated (by setting $x_{f(k)} = 1$ or 0) and denote with x_j^k ($j = r, \ldots, f(k)$) the sequence of values assigned to variables $x_r, \ldots, x_{f(k)}$ along the path in the tree from the root to k. The set of terminal nodes (*leaves*) of the tree can then be partitioned into

$$L_1 = \left\{ l : \sum_{j=r}^{f(l)} w_j x_j^l > c(r) \right\} \qquad \text{(infeasible leaves)},$$

$$L_2 = \left\{ l : f(l) = t \text{ and } \sum_{j=r}^{f(l)} w_j x_j^l \le c(r) \right\} \qquad \text{(feasible leaves)}.$$

For each $l \in L_1 \cup L_2$, let u_l be any upper bound on the problem defined by (2.1), (2.2) and

$$\begin{cases} x_j = x_j^l & \text{if } j \in \{r, \ldots, f(l)\}, \\ x_j = 0 \text{ or } 1 & \text{if } j \in N \backslash \{r, \ldots, f(l)\}. \end{cases} \tag{2.23}$$

Since all the nonleaf nodes are completely explored by the tree, a valid upper bound for KP is given by

$$U_6 = \max \{u_l : l \in L_1 \cup L_2\}. \tag{2.24}$$

A fast way to compute u_l is the following. Let $p^l = \sum_{j=1}^{r-1} p_j + \sum_{j=r}^{f(l)} p_j x_j^l$, and $d^l = | c(r) - \sum_{j=r}^{f(l)} w_j x_j^l |$; then

$$u_l = \begin{cases} \left\lfloor p^l - d^l \dfrac{p_{r-1}}{w_{r-1}} \right\rfloor & \text{if } l \in L_1; \\ \left\lfloor p^l + d^l \dfrac{p_{t+1}}{w_{t+1}} \right\rfloor & \text{if } l \in L_2, \end{cases} \tag{2.25}$$

which is clearly an upper bound on the continuous solution value for problem (2.1), (2.2), (2.23).

The computation of U_6 requires $O(n)$ time to determine the critical item and define KP(r, t), plus $O(2^{t-r})$ time to explore the binary tree. If $t - r$ is bounded by a constant, the overall time complexity is thus $O(n)$.

Example 2.1 (continued)

Assume $r = 4$ and $t = 6$. The reduced capacity is $c(r) = 60$. The binary tree is given in Figure 2.1. The leaf sets are $L_1 = \{2, 8\}$, $L_2 = \{4, 5, 9, 11, 12\}$. It follows that $U_6 = 280$, which is the optimal solution value. □

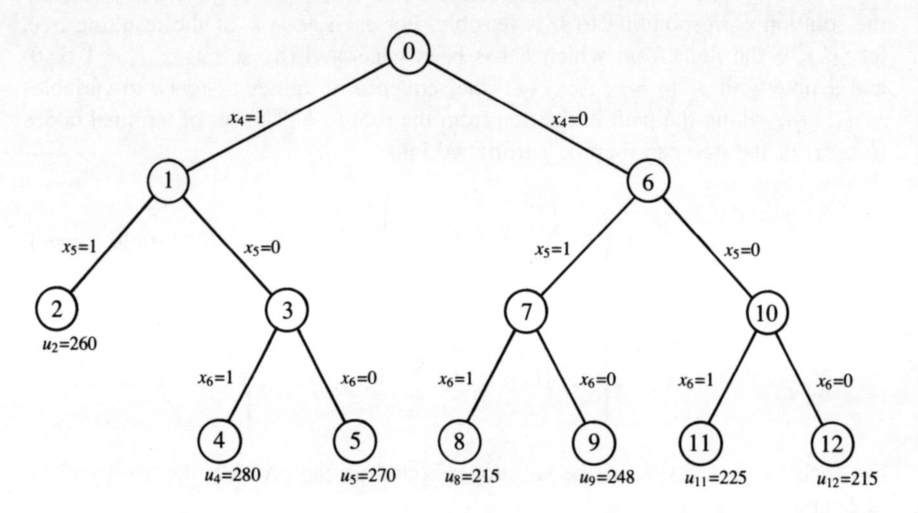

Figure 2.1 Binary tree for upper bound U_6 of Example 2.1

The upper bounds at the leaves can also be evaluated, of course, using any of the bounds previously described, instead of (2.25). If U_k ($k = 1, \dots, 5$) is used, then clearly $U_6 \le U_k$; if (2.25) is used, then no dominance exists between U_6 and the Dudzinsky and Walukiewicz (1984a) bound, so the best upper bound for KP is

$$U = \min\,(U_5, U_6).$$

U_6 can be strengthened, with very small extra computational effort, by evaluating $w_m = \min\{w_j : j > t\}$. It is not difficult to see that, when $l \in L_2$ and $d^l < w_m$, u_l can be computed as

$$u_l = \max\left(p^l, \left\lfloor p^l + w_m \frac{p_{t+1}}{w_{t+1}} - (w_m - d^l)\frac{p_{r-1}}{w_{r-1}}\right\rfloor\right). \tag{2.26}$$

Finally, we note that the computation of U_6 can be considerably accelerated by using an appropriate branch-and-bound algorithm to solve KP(r, t). At any iteration of such algorithm, let $\overline{z}(r, t)$ be the value of the best solution so far. For any nonleaf node k of the decision-tree, let $\overline{u}_k$ be an upper bound on the optimal solution of the subproblem defined by items $r, \dots, n$ with reduced capacity $c(r)$, i.e., the subproblem obtained by setting $x_j = 1$ for $j = 1, \dots, r-1$. $\overline{u}_k$ can be computed as an upper bound of the continuous solution value of the problem, i.e.

$$\overline{u}_k = \sum_{j=r}^{f(k)} p_j x_j^k + \sum_{j=f(k)+1}^{s(k)-1} p_j$$

$$+ \left\lfloor \left(c(r) - \left(\sum_{j=r}^{f(k)} w_j x_j^k + \sum_{j=f(k)+1}^{s(k)-1} w_j \right) \right) \frac{p_{s(k)}}{w_{s(k)}} \right\rfloor, \tag{2.27}$$

where $s(k) = \min(t+1, \min\{i : \sum_{j=r}^{f(k)} w_j x_j^k + \sum_{j=f(k)+1}^{i} w_j > c(r)\})$. If we have $\overline{u}_k \leq \overline{z}(r,t)$, the nodes descending from k need not be generated. In fact, for any leaf l descending from k, it would result that $u_l \leq \sum_{j=1}^{r-1} p_j + \overline{u}_k \leq \sum_{j=1}^{r-1} p_j + z(KP(r,t)) \leq U_6$.

Example 2.1 (continued)

Accelerating the computation through (2.27), we obtain the reduced branch-decision tree of Figure 2.2. □

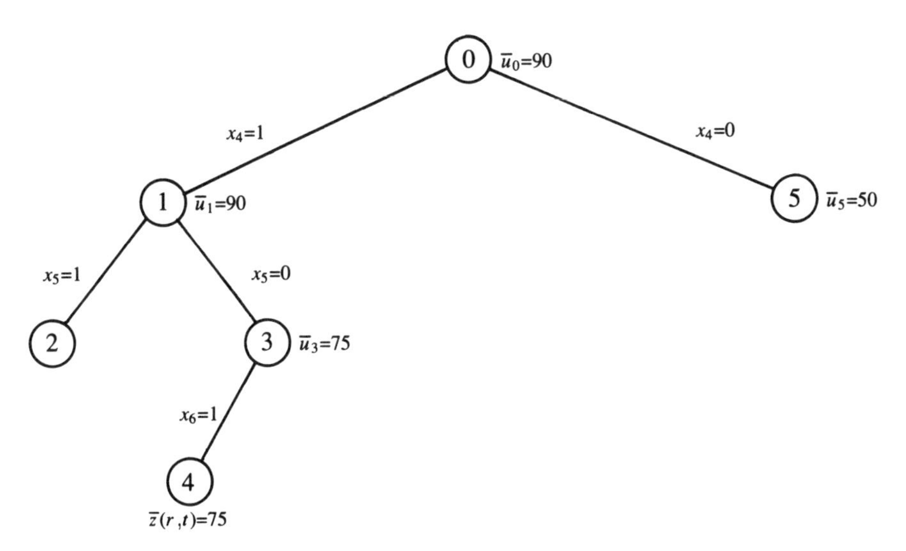

Figure 2.2 Branch-and-bound tree for upper bound U_6 of Example 2.1

2.4 THE GREEDY ALGORITHM

The most immediate way to determine an approximate solution to KP exploits the fact that solution $\overline{x}$ of the continuous relaxation of the problem has only one fractional variable, $\overline{x}_s$ (see Theorem 2.1). Setting this variable to 0 gives a feasible solution to KP of value

$$z' = \sum_{j=1}^{s-1} p_j .$$

We can expect that z' is, on average, quite close to the optimal solution value z. In fact $z' \le z \le U_1 \le z' + p_s$, i.e. the absolute error is bounded by p_s. The worst-case performance ratio, however, is arbitrarily bad. This is shown by the series of problems with $n = 2$, $p_1 = w_1 = 1$, $p_2 = w_2 = k$ and $c = k$, for which $z' = 1$ and $z = k$, so the ratio z'/z is arbitrarily close to 0 for k sufficiently large.

Noting that the above pathology occurs when p_s is relatively large, we can obtain an improved heuristic by also considering the feasible solution given by the critical item alone and taking the best of the two solution values, i.e.

$$z^h = \max\,(z', p_s). \tag{2.28}$$

The worst-case performance ratio of the new heuristic is $\frac{1}{2}$. We have already noted, in fact, that $z \le z' + p_s$, so, from (2.28), $z \le 2z^h$. To see that $\frac{1}{2}$ is tight, consider the series of problems with $n = 3$, $p_1 = w_1 = 1$, $p_2 = w_2 = p_3 = w_3 = k$ and $c = 2k$: we have $z^h = k + 1$ and $z = 2k$, so z^h/z is arbitrarily close to $\frac{1}{2}$ for k sufficiently large.

The computation of z^h requires $O(n)$ time, once the critical item is known. If the items are sorted as in (2.7), a more effective algorithm is to consider them according to increasing indices and insert each new item into the knapsack if it fits. (Notice that items $1, \ldots, s-1$ are always inserted, so the solution value is at least z'.) This is the most popular heuristic approach to KP, usually called the *Greedy Algorithm*. Again, the worst-case performance can be as bad as 0 (take for example the series of problems introduced for z'), but can be improved to $\frac{1}{2}$ if we also consider the solution given by the item of maximum profit alone, as in the following implementation. We assume that the items are ordered according to (2.7).

procedure GREEDY:
input: $n, c, (p_j), (w_j)$;
output: $z^g, (x_j)$;
begin
  $\overline{c} := c$;
  $z^g := 0$;
  $j^* := 1$;
  for $j := 1$ **to** n **do**
    begin
      if $w_j > \overline{c}$ **then** $x_j := 0$
      else
        begin
          $x_j := 1$;

```
                    c̄ := c̄ − w_j;
                    z^g := z^g + p_j
                end;
            if p_j > p_j* then j* := j
        end;
    if p_j* > z^g then
        begin
            z^g := p_j*;
            for j := 1 to n do x_j := 0;
            x_j* := 1
        end
end.
```

The worst-case performance ratio is $\frac{1}{2}$ since: (a) $p_{j^*} \geq p_s$, so $z^g \geq z^h$; (b) the series of problems introduced for z^h proves the tightness. The time complexity is $O(n)$, plus $O(n \log n)$ for the initial sorting.

For Example 2.1 we have $z' = z^h = 265$ and $z^g = 280$, which is the optimal solution value since $U_6 = 280$.

When a 0-1 knapsack problem in minimization form (see Section 2.1) is heuristically solved by applying GREEDY to its equivalent maximization instance, we of course obtain a feasible solution, but the worst-case performance is not preserved. Consider, in fact, the series of minimization problems with $n = 3$, $p_1 = w_1 = k$, $p_2 = w_2 = 1$, $p_3 = w_3 = k$ and $q = 1$, for which the optimal solution value is 1. Applying GREEDY to the maximization version (with $c = 2k$), we get $z^g = k + 1$ and hence an arbitrarily bad heuristic solution of value k for the minimization problem.

Other approximate algorithms for KP are considered in Section 2.8.

2.5 BRANCH-AND-BOUND ALGORITHMS

The first branch-and-bound approach to the exact solution of KP was presented by Kolesar (1967). His algorithm consists of a *highest-first* binary branching scheme which: (a) at each node, selects the not-yet-fixed item j having the maximum profit per unit weight, and generates two descendent nodes by fixing x_j, respectively, to 1 and 0; (b) continues the search from the feasible node for which the value of upper bound U_1 is a maximum.

The large computer memory and time requirements of the Kolesar algorithm were greatly reduced by the Greenberg and Hegerich (1970) approach, differing in two main respects: (a) at each node, the continuous relaxation of the induced subproblem is solved and the corresponding critical item $\tilde{s}$ is selected to generate the two descendent nodes (by imposing $x_{\tilde{s}} = 0$ and $x_{\tilde{s}} = 1$); (b) the search continues from the node associated with the exclusion of item $\tilde{s}$ (condition $x_{\tilde{s}} = 0$). When the continuous relaxation has an all-integer solution, the search is resumed from the last node generated by imposing $x_{\tilde{s}} = 1$, i.e. the algorithm is of *depth-first* type.

Horowitz and Sahni (1974) (and, independently, Ahrens and Finke (1975)) derived from the previous scheme a depth-first algorithm in which: (a) selection of the branching variable x_j is the same as in Kolesar; (b) the search continues from the node associated with the insertion of item j (condition $x_j = 1$), i.e. following a greedy strategy.

Other algorithms have been derived from the Greenberg–Hegerich approach (Barr and Ross (1975), Laurière (1978)) and from different techniques (Lageweg and Lenstra (1972), Guignard and Spielberg (1972), Fayard and Plateau (1975), Veliev and Mamedov (1981)). The Horowitz–Sahni one is, however, the most effective, structured and easy to implement, and has constituted the basis for several improvements.

2.5.1 The Horowitz–Sahni algorithm

Assume that the items are sorted as in (2.7). A *forward move* consists of inserting the largest possible set of new consecutive items into the current solution. A *backtracking move* consists of removing the last inserted item from the current solution. Whenever a forward move is exhausted, the upper bound U_1 corresponding to the current solution is computed and compared with the best solution so far, in order to check whether further forward moves could lead to a better one: if so, a new forward move is performed, otherwise a backtracking follows. When the last item has been considered, the current solution is complete and possible updating of the best solution so far occurs. The algorithm stops when no further backtracking can be performed.

In the following description of the algorithm we use the notations

$(\hat{x}_j)$ = current solution;

$\hat{z}$ = current solution value $\left(= \sum_{j=1}^{n} p_j \hat{x}_j \right)$;

$\hat{c}$ = current residual capacity $\left(= c - \sum_{j=1}^{n} w_j \hat{x}_j \right)$;

(x_j) = best solution so far;

z = value of the best solution so far $\left(= \sum_{j=1}^{n} p_j x_j \right)$.

procedure HS:
input: $n, c, (p_j), (w_j)$;
output: $z, (x_j)$;
begin
1. [initialize]
    $z := 0$;
    $\hat{z} := 0$;

```
        ĉ := c;
        p_{n+1} := 0;
        w_{n+1} := +∞;
        j := 1;
2. [compute upper bound U_1]
        find  r = min {i : Σ_{k=j}^{i} w_k > ĉ};
        u := Σ_{k=j}^{r-1} p_k + ⌊(ĉ − Σ_{k=j}^{r-1} w_k)p_r/w_r⌋;
        if z ≥ ẑ + u then go to 5;
3. [perform a forward step]
        while w_j ≤ ĉ do
            begin
                ĉ := ĉ − w_j;
                ẑ := ẑ + p_j;
                x̂_j := 1;
                j := j + 1
            end;
        if j ≤ n then
            begin
                x̂_j := 0 ;
                j := j + 1
            end;
        if j < n then go to 2;
        if j = n then go to 3;
4. [update the best solution so far]
        if ẑ > z then
            begin
                z := ẑ;
                for k := 1 to n do x_k := x̂_k
            end;
        j := n;
        if x̂_n = 1 then
            begin
                ĉ := ĉ + w_n;
                ẑ := ẑ − p_n;
                x̂_n := 0
            end;
5. [backtrack]
        find  i = max {k < j : x̂_k = 1};
        if no such i then return ;
        ĉ := ĉ + w_i;
        ẑ := ẑ − p_i;
        x̂_i := 0;
        j := i + 1;
        go to  2
end.
```

Example 2.2

Consider the instance of KP defined by

$n = 7;$

$(p_j) = (70, 20, 39, 37, 7, 5, 10);$

$(w_j) = (31, 10, 20, 19, 4, 3, 6);$

$c = 50.$

Figure 2.3 gives the decision-tree produced by procedure HS. □

Several effective algorithms have been obtained by improving the Horowitz–Sahni strategy. Mention is made in particular of those of Nauss (1976) (with Fortran code available), Martello and Toth (1977a) (with Fortran code in Martello and Toth (1978) and Pascal code in Syslo, Deo and Kowalik (1983)), Suhl (1978), Zoltners (1978).

We describe the Martello–Toth algorithm, which is generally considered highly effective.

2.5.2 The Martello–Toth algorithm

The method differs from that of Horowitz and Sahni (1974) in the following main respects (we use the notations introduced in the previous section).

(i) Upper bound U_2 is used instead of U_1.

(ii) The forward move associated with the selection of the jth item is split into two phases: *building of a new current solution* and *saving of the current solution.* In the first phase the largest set N_j of consecutive items which can be inserted into the current solution starting from the jth, is defined, and the upper bound corresponding to the insertion of the jth item is computed. If this bound is less than or equal to the value of the best solution so far, a backtracking move immediately follows. If it is greater, the second phase, that is, insertion of the items of set N_j into the current solution, is performed only if the value of such a new solution does not represent the maximum which can be obtained by inserting the jth item. Otherwise, the best solution so far is changed, but the current solution is not updated, so useless backtrackings on the items in N_j are avoided.

(iii) A particular forward procedure, based on dominance criteria, is performed whenever, before a backtracking move on the ith item, the residual capacity $\hat{c}$ does not allow insertion into the current solution of any item following the ith. The procedure is based on the following consideration: the current solution could be improved only if the ith item is replaced by an item having greater profit and a weight small enough to allow its insertion, or by at least two items having global weight not greater than $w_i + \hat{c}$. By this approach it is generally possible to eliminate most of the useless nodes generated at the lowest levels of the decision-tree.

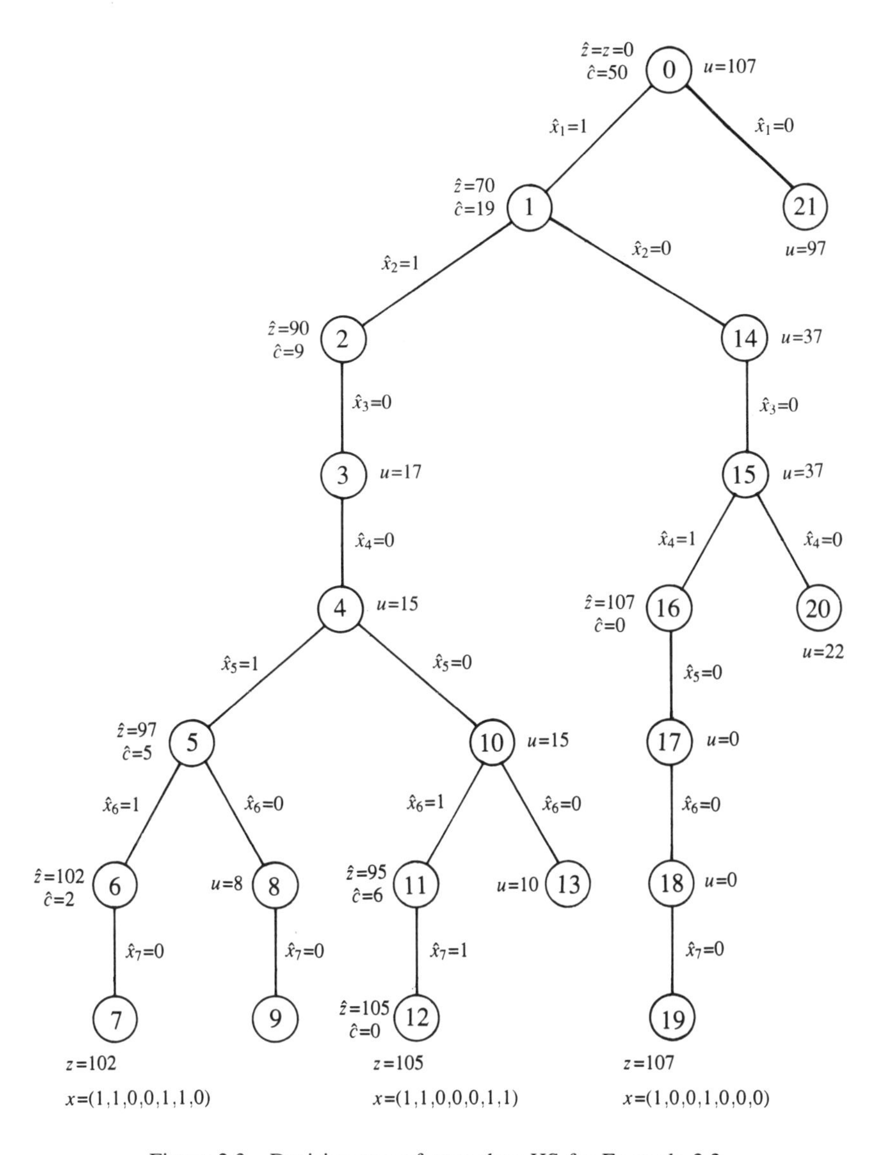

Figure 2.3 Decision-tree of procedure HS for Example 2.2

(iv) The upper bounds associated with the nodes of the decision-tree are computed through a parametric technique based on the storing of information related to the current solution. Suppose, in fact, that the current solution has been built by inserting all the items from the jth to the rth: then, when performing a backtracking on one of these items (say the ith, $j \leq i < r$), if no insertion occurred for the items preceding the jth, it is possible to insert at least items $i+1, \ldots, r$ into the new current solution. To this end, we store in $\overline{r}_i$, $\overline{p}_i$ and

$\overline{w}_i$ the quantities $r+1$, $\sum_{k=i}^{r} p_k$ and $\sum_{k=i}^{r} w_k$, respectively, for $i = j, \ldots, r$, and in $\tilde{r}$ the value $r-1$ (used for subsequent updatings).

Detailed description of the algorithm follows (it is assumed that the items are sorted as in (2.7)).

procedure MT1:
input: $n, c, (p_j), (w_j)$;
output: $z, (x_j)$;
begin
1. [initialize]
   $z := 0$;
   $\hat{z} := 0$;
   $\hat{c} := c$;
   $p_{n+1} := 0$;
   $w_{n+1} := +\infty$;
   for $k := 1$ **to** n **do** $\hat{x}_k := 0$;
   compute the upper bound $U = U_2$ on the optimal solution value;
   $\overline{w}_1 := 0$;
   $\overline{p}_1 := 0$;
   $\overline{r}_1 := 1$;
   $\tilde{r} := n$;
   for $k := n$ **to** 1 **step** -1 **do** compute $m_k = \min\{w_i : i > k\}$;
   $j := 1$;
2. [build a new current solution]
   while $w_j > \hat{c}$ **do**
      if $z \geq \hat{z} + \lfloor \hat{c} p_{j+1}/w_{j+1} \rfloor$ **then go to** 5 **else** $j := j+1$;
   find $r = \min\{i : \overline{w}_j + \sum_{k=\overline{r}_j}^{i} w_k > \hat{c}\}$;
   $p' := \overline{p}_j + \sum_{k=\overline{r}_j}^{r-1} p_k$;
   $w' := \overline{w}_j + \sum_{k=\overline{r}_j}^{r-1} w_k$;
   if $r \leq n$ **then** $u := \max(\lfloor (\hat{c} - w')p_{r+1}/w_{r+1} \rfloor,$
      $\lfloor p_r - (w_r - (\hat{c} - w'))p_{r-1}/w_{r-1} \rfloor)$
   else $u := 0$;
   if $z \geq \hat{z} + p' + u$ **then go to** 5;
   if $u = 0$ **then go to** 4;
3. [save the current solution]
   $\hat{c} := \hat{c} - w'$;
   $\hat{z} := \hat{z} + p'$;
   for $k := j$ **to** $r-1$ **do** $\hat{x}_k := 1$;
   $\overline{w}_j := w'$;
   $\overline{p}_j := p'$;
   $\overline{r}_j := r$;
   for $k := j+1$ **to** $r-1$ **do**
      begin
         $\overline{w}_k := \overline{w}_{k-1} - w_{k-1}$;
         $\overline{p}_k := \overline{p}_{k-1} - p_{k-1}$;
         $\overline{r}_k := r$

```
        end;
    for k := r to r̃ do
        begin
            w̄_k := 0;
            p̄_k := 0;
            r̄_k := k
        end;
    r̃ := r − 1;
    j := r + 1;
    if ĉ ≥ m_{j−1} then go to 2;
    if z ≥ ẑ then go to 5;
    p' := 0;
4. [update the best solution so far]
    z := ẑ + p';
    for k := 1 to j − 1 do x_k := x̂_k;
    for k := j to r − 1 do x_k := 1;
    for k := r to n do x_k := 0;
    if z = U then return ;
5. [backtrack]
    find i = max {k < j : x̂_k = 1};
    if no such i then return;
    ĉ := ĉ + w_i;
    ẑ := ẑ − p_i;
    x̂_i := 0;
    j := i + 1;
    if ĉ − w_i ≥ m_i then go to 2;
    j := i;
    h := i;
6. [try to replace item i with item h]
    h := h + 1;
    if z ≥ ẑ + ⌊ĉp_h/w_h⌋ then go to 5;
    if w_h = w_i then go to 6;
    if w_h > w_i then
        begin
            if w_h > ĉ or z ≥ ẑ + p_h then go to 6;
            z := ẑ + p_h;
            for k := 1 to n do x_k := x̂_k;
            x_h := 1;
            if z = U then return;
            i := h;
            go to 6
        end
    else
        begin
            if ĉ − w_h < m_h then go to 6;
            ĉ := ĉ − w_h;
            ẑ := ẑ + p_h;
            x̂_h := 1;
            j := h + 1;
```

```
            w̄_h := w_h;
            p̄_h := p_h;
            r̄_h := h + 1;
            for k := h + 1 to r̃ do
                begin
                    w̄_k := 0;
                    p̄_k := 0;
                    r̄_k := k
                end;
            r̃ := h;
            go to 2
        end
end.
```

The Fortran code corresponding to MT1 is included in the present volume. In addition, a second code, MT1R, is included which accepts on input real values for profits, weights and capacity.

Example 2.2 (continued)

Figure 2.4 gives the decision-tree produced by procedure MT1. □

Branch-and-bound algorithms are nowadays the most common way to effectively find the optimal solution of knapsack problems. More recent techniques imbed the branch-and-bound process into a particular algorithmic framework to solve, with increased efficiency, large instances of the problem. We describe them in Section 2.9.

The other fundamental approach to KP is dynamic programming. This has been the first technique available for exactly solving the problem and, although its importance has decreased in favour of branch-and-bound, it is still interesting because (a) it usually beats the other methods when the instance is very hard (see the computational results of Section 2.10.1), and (b) it can be successfully used in combination with branch-and-bound to produce hybrid algorithms for KP (Plateau and Elkihel, 1985) and for other knapsack-type problems (Martello and Toth (1984a), Section 4.2.2).

2.6 DYNAMIC PROGRAMMING ALGORITHMS

Given a pair of integers m ($1 \leq m \leq n$) and $\hat{c}$ ($0 \leq \hat{c} \leq c$), consider the sub-instance of KP consisting of items $1, \ldots, m$ and capacity $\hat{c}$. Let $f_m(\hat{c})$ denote its optimal solution value, i.e.

$$f_m(\hat{c}) = \max \left\{ \sum_{j=1}^{m} p_j x_j : \sum_{j=1}^{m} w_j x_j \leq \hat{c},\ x_j = 0 \text{ or } 1 \text{ for } j = 1, \ldots, m \right\}. \quad (2.29)$$

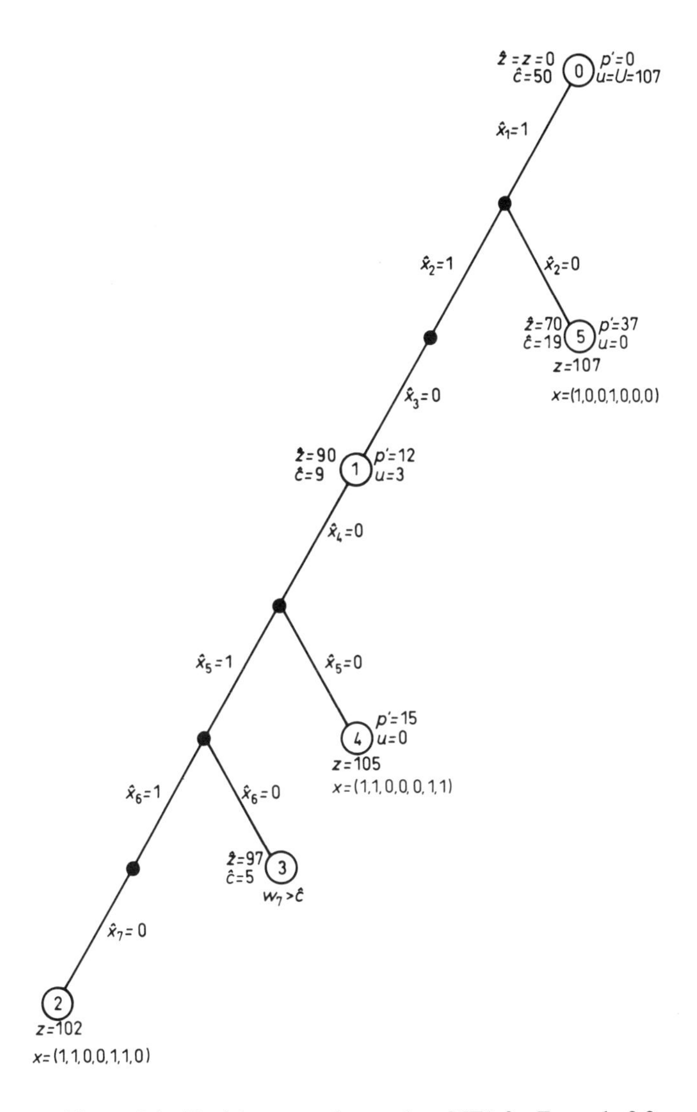

Figure 2.4 Decision-tree of procedure MT1 for Example 2.2

We trivially have

$$f_1(\hat{c}) = \begin{cases} 0 & \text{for } \hat{c} = 0, \ldots, w_1 - 1; \\ p_1 & \text{for } \hat{c} = w_1, \ldots, c. \end{cases}$$

Dynamic programming consists of considering n *stages* (for m increasing from 1 to n) and computing, at each stage $m > 1$, the values $f_m(\hat{c})$ (for $\hat{c}$ increasing from 0 to c) using the classical recursion (Bellman, 1954, 1957; Dantzig, 1957):

$$f_m(\hat{c}) = \begin{cases} f_{m-1}(\hat{c}) & \text{for } \hat{c} = 0, \ldots, w_m - 1; \\ \max\,(f_{m-1}(\hat{c}), f_{m-1}(\hat{c} - w_m) + p_m) & \text{for } \hat{c} = w_m, \ldots, c. \end{cases}$$

We call *states* the feasible solutions corresponding to the $f_m(\hat{c})$ values. The optimal solution of the problem is the state corresponding to $f_n(c)$.

Toth (1980) directly derived from the Bellman recursion an efficient procedure for computing the states of a stage. The following values are assumed to be defined before execution for stage m:

$$v = \min\left(\sum_{j=1}^{m-1} w_j, c\right); \tag{2.30}$$

$$b = 2^{m-1}; \tag{2.31}$$

$$P_{\hat{c}} = f_{m-1}(\hat{c}), \qquad \text{for } \hat{c} = 0, \ldots, v; \tag{2.32}$$

$$X_{\hat{c}} = \{x_{m-1}, x_{m-2}, \ldots, x_1\}, \qquad \text{for } \hat{c} = 0, \ldots, v, \tag{2.33}$$

where x_j defines the value of the jth variable in the partial optimal solution corresponding to $f_{m-1}(\hat{c})$, i.e.

$$\hat{c} = \sum_{j=1}^{m-1} w_j x_j \quad \text{and} \quad f_{m-1}(\hat{c}) = \sum_{j=1}^{m-1} p_j x_j .$$

From a computational point of view, it is convenient to encode each set $X_{\hat{c}}$ as a bit string, so this notation will be used in the following. After execution, values (2.30) to (2.33) are relative to stage m.

```
procedure REC1:
input: v, b, (P_ĉ), (X_ĉ), w_m, p_m;
output: v, b, (P_ĉ), (X_ĉ);
begin
    if v < c then
        begin
            u := v;
            v := min (v + w_m, c);
            for ĉ := u + 1 to v do
                begin
                    P_ĉ := P_u;
                    X_ĉ := X_u
                end
        end;
    for ĉ := v to w_m step −1 do
        if P_ĉ < P_{ĉ−w_m} + p_m then
            begin
```

```
                    P_ĉ := P_{ĉ-w_m} + p_m;
                    X_ĉ := X_{ĉ-w_m} + b
                end;
    b := 2b
end.
```

An immediate dynamic programming algorithm for KP is thus the following.

```
procedure DP1:
input: n, c, (p_j), (w_j);
output: z, (x_j);
begin
    for ĉ := 0 to w_1 - 1 do
        begin
            P_ĉ := 0;
            X_ĉ := 0
        end;
    v := w_1;
    b := 2;
    P_v := p_1;
    X_v := 1;
    for m := 2 to n do call REC1;
    z := P_c;
    determine (x_j) by decoding X_c
end.
```

Procedure REC1 requires $O(c)$ time, so the time complexity of DP1 is $O(nc)$. The space complexity is $O(nc)$. By encoding $X_{\hat{c}}$ as a bit string in computer words of d bits, the actual storage requirement is $(1 + \lceil n/d \rceil)c$, where $\lceil a \rceil$ is the smallest integer not less than a.

2.6.1 Elimination of dominated states

The number of states considered at each stage can be considerably reduced by eliminating *dominated states*, that is, those states $(P_{\hat{c}}, X_{\hat{c}})$ for which there exists a state (P_y, X_y) with $P_y \geq P_{\hat{c}}$ and $y < \hat{c}$. (Any solution obtainable from $(P_{\hat{c}}, X_{\hat{c}})$ can be obtained from (P_y, X_y).) This technique has been used by Horowitz and Sahni (1974) and Ahrens and Finke (1975). The undominated states of the mth stage can be computed through a procedure proposed by Toth (1980). The following values are assumed to be defined before execution of the procedure for stage m:

$$s = \text{number of states at stage } (m-1); \tag{2.34}$$

$$b = 2^{m-1}; \tag{2.35}$$

$$W1_i = \text{total weight of the } i\text{th state } (i = 1, \ldots, s); \tag{2.36}$$

$$P1_i = \text{total profit of the } i\text{th state } (i = 1, \ldots, s); \tag{2.37}$$

$$X1_i = \{x_{m-1}, x_{m-2}, \ldots, x_1\}, \quad \text{for } i = 1, \ldots, s, \tag{2.38}$$

where x_j defines the value of the jth variable in the partial optimal solution of the ith state, i.e.

$$W1_i = \sum_{j=1}^{m-1} w_j x_j \quad \text{and} \quad P1_i = \sum_{j=1}^{m-1} p_j x_j .$$

Vector $W1$ (and, hence, $P1$) is assumed to be ordered according to strictly increasing values.

The procedure uses index i to scan the states of the current stage and index k to store the states of the new stage. Each current state can produce a new state of total weight $y = W1_i + w_m$, so the current states of total weight $W1_h < y$, and then the new state, are stored in the new stage, but only if they are not dominated by a state already stored. After execution, values (2.34) and (2.35) are relative to the new stage, while the new values of (2.36), (2.37) and (2.38) are given by $(W2_k)$, $(P2_k)$ and $(X2_k)$, respectively. Sets $X1_i$ and $X2_k$ are encoded as bit strings. Vectors $(W2_k)$ and $(P2_k)$ result ordered according to strictly increasing values. On input, it is assumed that $W1_0 = P1_0 = X1_0 = 0$.

```
procedure REC2:
input: s, b, (W1_i), (P1_i), (X1_i), w_m, p_m, c;
output: s, b, (W2_k), (P2_k), (X2_k);
begin
    i := 0;
    k := 0;
    h := 1;
    y := w_m;
    W1_{s+1} := +∞;
    W2_0 := 0;
    P2_0 := 0;
    X2_0 := 0;
    while min (y, W1_h) ≤ c do
        if W1_h ≤ y then
            begin
                comment: define the next state (p, x);
                p := P1_h;
                x := X1_h;
                if W1_h = y then
                    begin
                        if P1_i + p_m > p then
                            begin
                                p := P1_i + p_m;
```

```
                                x := X1_i + b
                            end;
                        i := i + 1;
                        y := W1_i + w_m
                    end;
                comment: store the next state, if not dominated;
                if p > P2_k then
                    begin
                        k := k + 1;
                        W2_k := W1_h;
                        P2_k := p;
                        X2_k := x
                    end;
                h := h + 1
            end
        else
            begin
                comment: store the new state, if not dominated;
                if P1_i + p_m > P2_k then
                    begin
                        k := k + 1;
                        W2_k := y;
                        P2_k := P1_i + p_m;
                        X2_k := X1_i + b
                    end;
                i := i + 1;
                y := W1_i + w_m
            end;
    s := k;
    b := 2b
end.
```

A dynamic programming algorithm using REC2 to solve KP is the following.

```
procedure DP2:
input: n, c, (p_j), (w_j);
output: z, (x_j);
begin
    W1_0 := 0;
    P1_0 := 0;
    X1_0 := 0;
    s := 1;
    b := 2;
    W1_1 := w_1;
    P1_1 := p_1;
    X1_1 := 1;
    for m := 2 to n do
        begin
```

```
            call REC2;
            rename W2, P2 and X2 as W1, P1 and X1, respectively
        end;
    z := P1_s;
    determine (x_j) by decoding X1_s
end.
```

The time complexity of REC2 is $O(s)$. Since s is bounded by min $(2^m - 1, c)$, the time complexity of DP2 is $O(\min(2^{n+1}, nc))$.

Procedure DP2 requires no specific ordering of the items. Its efficiency, however, improves considerably if they are sorted according to decreasing p_j/w_j ratios since, in this case, the number of undominated states is reduced. Hence, this ordering is assumed in the following.

Example 2.3

Consider the instance of KP defined by

$$n = 6;$$

$$(p_j) = (50, 50, 64, 46, 50, 5);$$

$$(w_j) = (56, 59, 80, 64, 75, 17);$$

$$c = 190.$$

Figure 2.5 gives, for each stage m and for each undominated state i, the values W_i, P_i, corresponding, in DP2, alternatively to $W1_i, P1_i$ and $W2_i, P2_i$. The optimal solution, of value 150, is $(x_j) = (1, 1, 0, 0, 1, 0)$. For the same example, procedure DP1 generates 866 states. □

	$m = 1$		$m = 2$		$m = 3$		$m = 4$		$m = 5$		$m = 6$	
i	W_i	P_i	W_i	P_i	W_i	P_i	W_i	P_i	W_i	P_i	W_i	P_i
0	0	0	0	0	0	0	0	0	0	0	0	0
1	56	50	56	50	56	50	56	50	56	50	17	5
2			115	100	80	64	80	64	80	64	56	50
3					115	100	115	100	115	100	73	55
4					136	114	136	114	136	114	80	64
5							179	146	179	146	97	69
6									190	150	115	100
7											132	105
8											136	114
9											153	119
10											179	146
11											190	150

Figure 2.5 States of procedure DP2 for Example 2.3

2.6.2 The Horowitz–Sahni algorithm

Horowitz and Sahni (1974) presented an algorithm based on the subdivision of the original problem of n variables into two subproblems, respectively of $q = \lceil n/2 \rceil$ and $r = n - q$ variables. For each subproblem a list containing all the undominated states relative to the last stage is computed; the two lists are then combined in order to find the optimal solution.

The main feature of the algorithm is the need, in the worst case, for two lists of $2^q - 1$ states each, instead of a single list of $2^n - 1$ states. Hence the time and space complexities decrease to $O(\min(2^{n/2}, nc))$, with a square root improvement in the most favourable case. In many cases, however, the number of undominated states is much lower than $2^{n/2}$, since (a) many states are dominated and (b) for n sufficiently large, we have, in general, $c \ll 2^{n/2}$.

Ahrens and Finke (1975) proposed an algorithm where the technique of Horowitz and Sahni is combined with a branch-and-bound procedure in order to further reduce storage requirements. The algorithm works very well for hard problems having low values of n and very high values of w_i and c, but has the disadvantage of always executing the branch-and-bound procedure, even when the storage requirements are not excessive.

We illustrate the Horowitz–Sahni algorithm with a numerical example.

Example 2.3 (continued)

We have $q = 3$. The algorithm generates the first list for m =1, 2, 3, and the second for m = 4, 5, 6. The corresponding undominated states are given in Figure 2.6. Combining the lists corresponding to m = 3 and m = 6 we get the final list of Figure 2.5. □

	$m = 1$		$m = 2$		$m = 3$		$m = 6$		$m = 5$		$m = 4$	
i	W_i	P_i	W_i	P_i	W_i	P_i	W_i	P_i	W_i	P_i	W_i	P_i
0	0	0	0	0	0	0	0	0	0	0	0	0
1	56	50	56	50	56	50	17	5	64	46	64	46
2			115	100	80	64	64	46	75	50		
3					115	100	75	50	139	96		
4					136	114	81	51				
5							92	55				
6							139	96				
7							156	101				

Figure 2.6 States of the Horowitz–Sahni algorithm for Example 2.3

2.6.3 The Toth algorithm

Toth (1980) presented a dynamic programming algorithm based on (a) the elimination of useless states and (b) a combination of procedures REC1 and REC2.

Several states computed by REC1 or REC2 are of no use for the following stages since, of course, we are only interested in states capable of producing, at the final stage, the optimal solution. Useless states produced by REC1 can be eliminated by the following rule:

If a state, defined at the mth stage, has a total weight W satisfying one of the conditions

$$\text{(i)}\ \ W < c - \sum_{j=m+1}^{n} w_j\,;$$

$$\text{(ii)}\ \ c - \min_{m<j\leq n}\{w_j\} < W < c,$$

then the state will never produce P_c and, hence, can be eliminated.

A similar rule can be given for REC2 (in this case, however, it is necessary to keep the largest-weight state satisfying (i)), and the last, i.e. sth, state. The rule cannot be extended, instead, to the Horowitz–Sahni algorithm, since, in order to combine the two lists, all the undominated states relative to the two subproblems must be known.

Example 2.3 (continued)

The states generated by DP2, with REC2 improved through the above elimination rule, are given in Figure 2.7. □

	$m = 1$		$m = 2$		$m = 3$		$m = 4$		$m = 5$		$m = 6$	
i	W_i	P_i	W_i	P_i	W_i	P_i	W_i	P_i	W_i	P_i	W_i	P_i
0	0	0	0	0	0	0	0	0	0	0	0	0
1	56	50	56	50	56	50	80	64	136	114	190	150
2			115	100	80	64	115	100	190	150		
3					115	100	136	114				
4					136	114	179	146				

Figure 2.7 States of the improved version of DP2 for Example 2.3

Algorithm DP2 is generally more efficient than DP1, because of the fewer number of states produced. Notice however that, for the computation of a single state, the time and space requirements of DP2 are higher. So, for hard problems, where very few states are dominated, and hence the two algorithms generate almost the same lists, DP1 must be preferred to DP2. A dynamic programming algorithm which effectively solves both easy and hard problems can thus be obtained by combining the best characteristics of the two approaches. This is achieved by using procedure REC2 as long as the number of generated states is low, and then passing to REC1. Simple heuristic rules to determine the iteration at which the procedure must be changed can be found in Toth (1980).

2.7 REDUCTION ALGORITHMS

The size of an instance of KP can be reduced by applying procedures to fix the optimal value of as many variables as possible. These procedures partition set $N = \{1, 2, \dots, n\}$ into three subsets:

$$J1 = \{j \in N : x_j = 1 \text{ in any optimal solution to KP}\},$$

$$J0 = \{j \in N : x_j = 0 \text{ in any optimal solution to KP}\},$$

$$F = N \backslash (J1 \cup J0).$$

The original KP can then be transformed into the reduced form

$$\text{maximize} \quad z = \sum_{j \in F} p_j x_j + \hat{p}$$

$$\text{subject to} \quad \sum_{j \in F} w_j x_j \leq \hat{c},$$

$$x_j = 0 \text{ or } 1, \quad j \in F,$$

where $\hat{p} = \sum_{j \in J1} p_j$, $\hat{c} = c - \sum_{j \in J1} w_j$.

Ingargiola and Korsh (1973) proposed the first method for determining $J1$ and $J0$. The basic idea is the following. If setting a variable x_j to a given value b ($b = 0$ or 1) produces infeasibility or implies a solution worse than an existing one, then x_j must take the value $(1 - b)$ in any optimal solution. Let l be the value of a feasible solution to KP, and, for $j \in N$, let u_j^1 (resp. u_j^0) be an upper bound for KP with the additional constraint $x_j = 1$ (resp. $x_j = 0$). Then we have

$$J1 = \{j \in N : u_j^0 < l\}, \tag{2.39}$$

$$J0 = \{j \in N : u_j^1 < l\}. \tag{2.40}$$

In the Ingargiola–Korsh algorithm, u_j^1 and u_j^0 are computed using the Dantzig bound. Let s be the critical item (see Section 2.2.1) and U_1 the Dantzig bound for the original problem. Then $u_j^1 = U_1$ for any $j < s$ and $u_j^0 = U_1$ for any $j > s$. Hence values $j > s$ (resp. $j < s$) need not be considered in determining $J1$ (resp. $J0$), since $U_1 \geq l$. The algorithm initializes l to $\sum_{j=1}^{s-1} p_j$ and improves it during execution. It is assumed that the items are ordered according to (2.7). Remember that $\sigma^1(j)$ and $\sigma^0(j)$ represent the critical item when it is imposed, respectively,

$x_j = 1$ and $x_j = 0$ (see (2.17) and (2.18)).

procedure IKR:
input: $n, c, (p_j), (w_j)$;
output: $J1, J0$;
begin
  $J1 := \emptyset$;
  $J0 := \emptyset$;
  determine $s = \min \{j : \sum_{i=1}^{j} w_i > c\}$;
  $l := \sum_{j=1}^{s-1} p_j$;
  for $j := 1$ **to** s **do**
    begin
      determine $\sigma^0(j)$ and compute u_j^0;
      $l := \max (l, \sum_{i=1, i \neq j}^{\sigma^0(j)-1} p_i)$;
      if $u_j^0 < l$ **then** $J1 := J1 \cup \{j\}$
    end;
  for $j := s$ **to** n **do**
    begin
      determine $\sigma^1(j)$ and compute u_j^1;
      $l := \max (l, p_j + \sum_{i=1}^{\sigma^1(j)-1} p_i)$;
      if $u_j^1 < l$ **then** $J0 := J0 \cup \{j\}$
    end
end.

Notice that the variables corresponding to items in $J1$ and $J0$ must take the fixed value in *any* optimal solution to KP, thus including the solution of value l when this is optimal. However, given a feasible solution $\tilde{x}$ of value l, we are only interested in finding a *better* one. Hence stronger definitions of $J1$ and $J0$ are obtained by replacing strict inequalities with inequalities in (2.39), (2.40), i.e.

$$J1 = \{j \in N : u_j^0 \leq l\}, \tag{2.41}$$

$$J0 = \{j \in N : u_j^1 \leq l\}. \tag{2.42}$$

If it turns out that the reduced problem is infeasible or has an optimal solution less than l, then $\tilde{x}$ is the optimal solution to the original problem.

Example 2.4

We use the same instance as in Example 2.2, whose optimal solution, of value 107, is $x = (1, 0, 0, 1, 0, 0, 0)$:

$$
\begin{aligned}
n &= 7; \\
(p_j) &= (70,\ 20,\ 39,\ 37,\ 7,\ 5,\ 10); \\
(w_j) &= (31,\ 10,\ 20,\ 19,\ 4,\ 3,\ 6); \\
c &= 50.
\end{aligned}
$$

Applying procedure IKR we get:

$s = 3, \quad l = 90;$

$j = 1: \quad u_1^0 = 97, \quad l = 96;$

$j = 2: \quad u_2^0 = 107;$

$j = 3: \quad u_3^0 = 107;$

$j = 3: \quad u_3^1 = 106;$

$j = 4: \quad u_4^1 = 107, \quad l = 107;$

$j = 5: \quad u_5^1 = 106;$

$j = 6: \quad u_6^1 = 106;$

$j = 7: \quad u_7^1 = 105,$

so $J1 = \emptyset,\ J0 = \{5, 6, 7\}$. □

In order to use definitions (2.41), (2.42) it is simply necessary to replace the $<$ sign with $\leq$ in the two tests of procedure IKR. With this modification we get $J1 = \emptyset,\ J0 = \{4, 5, 6, 7\}$. The optimal solution value of the reduced problem is then 90, implying that the feasible solution of value $l = 107$ is optimal. (Notice that it is worth storing the solution vector corresponding to l during execution.)

Recently, Murphy (1986) erroneously claimed that definitions (2.41), (2.42) of $J1$ and $J0$ are incorrect. Balas, Nauss and Zemel (1987) have pointed out its mistake.

The time complexity of the Ingargiola–Korsh procedure is $O(n^2)$, since $O(n)$ time is required for each $\sigma^0(j)$ or $\sigma^1(j)$ computation (although one can expect that, on average, these values can be determined with few operations, starting from s). The time complexity does not change if u_j^0 and u_j^1 are computed through one of the improved upper bounding techniques of Section 2.3.

An $O(n)$ reduction algorithm has been independently obtained by Fayard and Plateau (1975) and Dembo and Hammer (1980). The method, FPDHR, computes u_j^0 and u_j^1 through the values $p_j^* = p_j - w_j p_s / w_s$ (see (2.13)). Recalling that $|\ p_j^*\ |$ represents a lower bound on the decrease of $z(C(KP))$ corresponding to the change of the jth variable from $\overline{x}_j$ to $1 - \overline{x}_j$, we have

$$u_j^0 = \lfloor z(C(KP)) - p_j^* \rfloor, \qquad j = 1, \dots, s;$$

$$u_j^1 = \lfloor z(C(KP)) + p_j^* \rfloor, \qquad j = s, \dots, n,$$

which are computed in constant time, once $z(C(KP))$ is known. It is easy to see that the values u_j^0 and u_j^1 obtained in this way are not lower than those of procedure IKR, so the method is generally less effective, in the sense that the resulting sets $J0$ and $J1$ have smaller cardinality.

$O(n^2)$ reduction algorithms more effective than the Ingargiola–Korsh method have been obtained by Toth (1976), Laurière (1978) and Fayard and Plateau (1982).

An effective reduction method, still dominating the Ingargiola–Korsh one, but requiring $O(n \log n)$ time, has been proposed by Martello and Toth (1988). The algorithm differs from procedure IKR in the following main respects:

(a) u_j^0 and u_j^1 are computed through the stronger bound U_2;

(b) $J1$ and $J0$ are determined at the end, thus using the best heuristic solution found;

(c) at each iteration, upper bound and improved heuristic solution value are computed in $O(\log n)$ time by initially defining $\overline{w}_j = \sum_{i=1}^j w_i$ and $\overline{p}_j = \sum_{i=1}^j p_i$ $(j = 1, \dots, n)$ and then determining, through binary search, the current critical item $\overline{s}$ (i.e. $\sigma^0(j)$ or $\sigma^1(j)$).

The procedure assumes that the items are ordered according to (2.7) and that $p_j/w_j = -\infty$ if $j < 1$, $p_j/w_j = +\infty$ if $j > n$.

```
procedure MTR:
input: n, c, (p_j), (w_j);
output: J1, J0, l;
begin
    for j := 0 to n do compute p̄_j = Σ_{i=1}^j p_i and w̄_j = Σ_{i=1}^j w_i;
    find, through binary search, s such that w̄_{s-1} ≤ c < w̄_s;
    l := p̄_{s-1};
    c̄ := c - w̄_{s-1};
    for j := s + 1 to n do
        if w_j ≤ c̄ then
            begin
                l := l + p_j;
                c̄ := c̄ - w_j
            end;
    for j := 1 to s do
        begin
            find, through binary search, s̄ such that
                w̄_{s̄-1} ≤ c + w_j < w̄_s̄;
            c̄ := c + w_j - w̄_{s̄-1};
```

```
                u_j^0 := p̄_{s̄-1} - p_j + max (⌊c̄ p_{s̄+1}/w_{s̄+1}⌋,
                                        ⌊p_s̄ - (w_s̄ - c̄) p_{s̄-1}/w_{s̄-1}⌋);
                l := max (l, p̄_{s̄-1} - p_j)
            end;
        for j := s to n do
            begin
                find, through binary search, s̄ such that
                    w̄_{s̄-1} ≤ c - w_j < w̄_s̄;
                c̄ := c - w_j - w̄_{s̄-1};
                u_j^1 := p̄_{s̄-1} + p_j + max (⌊c̄ p_{s̄+1}/w_{s̄+1}⌋, ⌊p_s̄ - (w_s̄ - c̄) p_{s̄-1}/w_{s̄-1}⌋);
                l := max (l, p̄_{s̄-1} + p_j)
            end;
        J1 := {j ≤ s : u_j^0 ≤ l};
        J0 := {j ≥ s : u_j^1 ≤ l}
end.
```

Example 2.4 (continued)

Applying procedure MTR we have

$(\overline{p}_j) = (0, 70, 90, 129, 166, 173, 178, 188)$;

$(\overline{w}_j) = (0, 31, 41, 61, 80, 84, 87, 93)$;

$s = 3,\ l = 90,\ \overline{c} = 9$;

$l = 102,\ \overline{c} = 2$;

$j = 1: \quad \overline{s} = 5, \overline{c} = 1, u_1^0 = 97$;

$j = 2: \quad \overline{s} = 3, \overline{c} = 19, u_2^0 = 107$;

$j = 3: \quad \overline{s} = 4, \overline{c} = 9, u_3^0 = 107$;

$j = 3: \quad \overline{s} = 1, \overline{c} = 30, u_3^1 = 99$;

$j = 4: \quad \overline{s} = 2, \overline{c} = 0, u_4^1 = 107, l = 107$;

$j = 5: \quad \overline{s} = 3, \overline{c} = 5, u_5^1 = 106$;

$j = 6: \quad \overline{s} = 3, \overline{c} = 6, u_6^1 = 106$;

$j = 7: \quad \overline{s} = 3, \overline{c} = 3, u_7^1 = 105$;

$J1 = \{1, 2, 3\}, \quad J0 = \{3, 4, 5, 6, 7\}$.

The reduced problem is infeasible (x_3 is fixed both to 1 and to 0 and, in addition, $\sum_{j \in J1} w_j > c$), so the feasible solution of value 107 is optimal. □

Procedure MTR computes the initial value of l through the greedy algorithm. Any other heuristic, requiring no more than $O(n \log n)$ time, could be used with no time complexity alteration.

The number of fixed variables can be further increased by imposing conditions (2.5), (2.6) to the reduced problem, i.e. setting $J0 = J0 \cup \{j \in F : w_j >$

$c - \sum_{j \in J1} w_j\}$ and, if $\sum_{j \in F} w_j \leq c - \sum_{j \in J1} w_j$, $J1 = J1 \cup F$. In addition, the procedure can be re-executed for the items in F (since the values of u_j^0 and u_j^1 relative to the reduced problem can decrease) until no further variable is fixed. This, however, would increase the time complexity by a factor n, unless the number of re-executions is bounded by a constant.

2.8 APPROXIMATE ALGORITHMS

In Section 2.4 we have described the greedy algorithm, which provides an approximate solution to KP with worst-case performance ratio equal to $\frac{1}{2}$, in time $O(n)$ plus $O(n \log n)$ for the initial sorting. Better accuracy can be obtained through approximation schemes, which allow one to obtain any prefixed performance ratio. In this section we examine polynomial-time and fully polynomial-time approximation schemes for KP. Besides these deterministic results, the probabilistic behaviour of some approximate algorithms has been investigated. A thorough analysis of probabilistic aspects is outwith the scope of this book. The main results are outlined in Section 2.8.3 and, for the subset-sum problem, in Section 4.3.4. (The contents of such sections are based on Karp, Lenstra, McDiarmid and Rinnooy Kan (1985).)

2.8.1 Polynomial-time approximation schemes

The first approximation scheme for KP was proposed by Sahni (1975) and makes use of a greedy-type procedure which finds a heuristic solution by filling, in order of decreasing p_j/w_j ratios, that part of c which is left vacant after the items of a given set M have been put into the knapsack. Given $M \subset N$ and assuming that the items are sorted according to (2.7), the procedure is as follows.

procedure GS:
input: $n, c, (p_j), (w_j), M$;
output: z^g, X;
begin
    $z^g := 0$;
    $\hat{c} := c - \sum_{j \in M} w_j$;
    $X := \emptyset$;
    for $j := 1$ **to** n **do**
        if $j \notin M$ and $w_j \leq \hat{c}$ **then**
            begin
                $z^g := z^g + p_j$;
                $\hat{c} := \hat{c} - w_j$;
                $X := X \cup \{j\}$
            end
end.

Given a non-negative integer parameter k, the Sahni scheme S(k) is

procedure S(k):
input: $n, c, (p_j), (w_j)$;
output: z^h, X^h;
begin
  $z^h := 0$;
  for each $M \subset \{1, \dots, n\}$ such that $|M| \le k$ and $\sum_{j \in M} w_j \le c$ **do**
    begin
      call GS;
      if $z^g + \sum_{j \in M} p_j > z^h$ **then**
        begin
          $z^h := z^g + \sum_{j \in M} p_j$;
          $X^h := X \cup M$
        end
    end
end.

Since the time complexity of procedure GS is $O(n)$ and the number of times it is executed is $O(n^k)$, the time complexity of S(k) is $O(n^{k+1})$. The space complexity is $O(n)$.

Theorem 2.3 (Sahni, 1975) *The worst-case performance ratio of* S(k) *is* $r(\mathrm{S}(k)) = k/(k+1)$.

Proof. (a) Let Y be the set of items inserted into the knapsack in the optimal solution. If $|Y| \le k$, then S(k) gives the optimum, since all combinations of size $|Y|$ are tried. Hence, assume $|Y| > k$. Let $\overline{M}$ be the set of the k items of highest profit in Y, and denote the remaining items of Y with $j_1, \dots, j_r$, assuming $p_{j_i}/w_{j_i} \ge p_{j_{i+1}}/w_{j_{i+1}}$ $(i = 1, \dots, r-1)$. Hence, if z is the optimal solution value, we have

$$p_{j_i} \le \frac{z}{k+1} \qquad \text{for } i = 1, \dots, r. \tag{2.43}$$

Consider now the iteration of S(k) in which $M = \overline{M}$, and let j_m be the first item of $\{j_1, \dots, j_r\}$ not inserted into the knapsack by GS. If no such item exists then the heuristic solution is optimal. Otherwise we can write z as

$$z = \sum_{i \in \overline{M}} p_i + \sum_{i=1}^{m-1} p_{j_i} + \sum_{i=m}^{r} p_{j_i}, \tag{2.44}$$

while for the heuristic solution value returned by GS we have

$$z^g \ge \sum_{i \in \overline{M}} p_i + \sum_{i=1}^{m-1} p_{j_i} + \sum_{i \in Q} p_i, \tag{2.45}$$

where Q denotes the set of those items of $N\backslash\overline{M}$ which are in the heuristic solution but not in $\{j_1, \dots, j_r\}$ and whose index is less than j_m. Let $c^* = c - \sum_{i \in \overline{M}} w_i - \sum_{i=1}^{m-1} w_{j_i}$ and $\overline{c} = c^* - \sum_{i \in Q} w_i$ be the residual capacities available, respectively, in the optimal and the heuristic solution for the items of $N\backslash\overline{M}$ following j_{m-1}. Hence, from (2.44),

$$z \leq \sum_{i \in \overline{M}} p_i + \sum_{i=1}^{m-1} p_{j_i} + c^* \frac{p_{j_m}}{w_{j_m}};$$

by definition of m we have $\overline{c} < w_{j_m}$ and $p_i/w_i \geq p_{j_m}/w_{j_m}$ for $i \in Q$, so

$$z < \sum_{i \in \overline{M}} p_i + \sum_{i=1}^{m-1} p_{j_i} + p_{j_m} + \sum_{i \in Q} p_i.$$

Hence, from (2.45), $z < z^g + p_{j_m}$ and, from (2.43),

$$\frac{z^g}{z} > \frac{k}{k+1}.$$

(b) To prove that the bound is tight, consider the series of instances with: $n = k+2$; $p_1 = 2$, $w_1 = 1$; $p_j = w_j = L > 2$ for $j = 2, \dots, k+2$; $c = (k+1)L$. The optimal solution value is $z = (k+1)L$, while S(k) gives $z^h = kL + 2$. Hence, for L sufficiently large, the ratio z^k/z is arbitrarily close to $k/(k+1)$. □

Let $\overline{M}$ denote the maximum cardinality subset of $\{1, \dots, n\}$ such that $\sum_{j \in \overline{M}} w_j \leq c$. Then, clearly, for any $k \geq |\overline{M}|$, S(k) gives the optimal solution.

Example 2.5

Consider the instance of KP defined by

$n = 8;$

$(p_j) = (350, 400, 450, 20, 70, 8, 5, 5);$

$(w_j) = (\ 25,\ \ 35,\ \ 45,\ \ 5, 25, 3, 2, 2);$

$c = 104.$

The optimal solution $X = \{1, 3, 4, 5, 7, 8\}$ has value $z = 900$.

Applying S(k) with $k = 0$, we get the greedy solution: $X^h = \{1, 2, 4, 5, 6, 7, 8\}$, $z^h = 858$.

Applying S(k) with $k = 1$, we re-obtain the greedy solution for $M = \{1\}, \{2\}, \{4\}, \{5\}, \{6\}, \{7\}, \{8\}$. For $M = \{3\}$, we obtain $X^h = \{1, 3, 4, 5, 6\}$, $z^h = 898$.

Applying S(k) with $k = 2$, we obtain the optimal solution when $M = \{3, 7\}$. □

The Sahni algorithm is a *polynomial-time approximation scheme*, in the sense that any prefixed worst-case performance ratio can be obtained in a time bounded by a polynomial. However, the degree of the polynomial increases with k, so the time complexity of the algorithm is exponential in k, i.e. in the inverse of the *worst-case relative error* $\varepsilon = 1 - r$.

2.8.2 Fully polynomial-time approximation schemes

Ibarra and Kim (1975) have obtained a *fully polynomial-time approximation scheme*, i.e. a parametric algorithm which allows one to obtain any worst-case relative error (note that imposing ε is equivalent to imposing r) in polynomial time and space, and such that the time and space complexities grow polynomially also with the inverse of the worst-case relative error ε. The basic ideas in the Ibarra–Kim algorithm are: (a) to separate items according to profits into a class of "large" items and one of "small" items; (b) to solve the problem for the large items only, with profits scaled by a suitable scale factor δ, through dynamic programming. The dynamic programming list is stored in a table T of length $\lfloor (3/\varepsilon)^2 \rfloor + 1$; $T(k)$ = "*undefined*" or is of the form $(L(k), P(k), W(k))$, where $L(k)$ is a subset of $\{1, \dots, n\}$, $P(k) = \sum_{j \in L(k)} p_j$, $W(k) = \sum_{j \in L(k)} w_j$ and $k = \sum_{j \in L(k)} \overline{p}_j$ with $\overline{p}_j = \lfloor p_j / \delta \rfloor$. It is assumed that the items are ordered according to (2.7) and that the "small" items are inserted in set S preserving this order.

procedure IK(ε) :
input: $n, c, (p_j), (w_j)$;
output: z^h, X^h;
begin
  find the critical item s (see Section 2.2.1);
  if $\sum_{j=1}^{s-1} w_j = c$ **then**
    begin
      $z^h := \sum_{j=1}^{s-1} p_j$;
      $X^h := \{1, \dots, s-1\}$;
      return
    end;
  $\tilde{z} := \sum_{j=1}^{s} p_j$
  comment: $\tilde{z}/2 \le z < \tilde{z}$, since $z \ge \max(\sum_{j=1}^{s-1} p_j, p_s)$;
  $\delta := \tilde{z}(\varepsilon/3)^2$;
  $S := \emptyset$;
  $T(0) := (L(0), P(0), W(0)) := (\emptyset, 0, 0)$;
  $q := \lfloor \tilde{z}/\delta \rfloor$ (**comment**: $q = \lfloor (3/\varepsilon)^2 \rfloor$);
  comment: dynamic programming phase;
  for $i := 1$ **to** q **do** $T(i)$:= "*undefined*";
  for $j := 1$ **to** n **do**
    if $p_j \le \varepsilon\tilde{z}/3$ **then** $S := S \cup \{j\}$
    else

```
            begin
                p̄_j := ⌊p_j / δ⌋;
                for i := q − p̄_j to 0 step −1 do
                    if T(i) ≠ "undefined" and W(i) + w_j ≤ c then
                        if T(i + p̄_j) = "undefined"
                          or W(i + p̄_j) > W(i) + w_j then
                            T(i + p̄_j) := (L(i) ∪ {j}, P(i) + p_j, W(i) + w_j)
            end;
    comment: greedy phase;
    z^h := 0;
    for i := 0 to q do
        if T(i) ≠ "undefined" then
            begin
                z̄ := P(i) + Σ_{j∈A} p_j, where A is obtained by filling the residual
                    capacity c − W(i) with items of S in the greedy way;
                if z̄ > z^h then
                    begin
                        z^h := z̄;
                        X^h := L(i) ∪ A
                    end
            end
end.
```

The dynamic programming recursion is executed n times and, at each iteration, no more than q states are considered: since each state takes a constant amount of time, the dynamic programming phase has time complexity $O(nq)$. The final greedy phase is performed at most q times, each iteration taking $O(n)$ time. Hence the overall time complexity of IK(ε) is $O(nq)$, i.e. $O(n/\varepsilon^2)$ by definition of q, plus $O(n \log n)$ for the initial sorting.

The space required by the algorithm is determined by the $\lfloor (3/\varepsilon)^2 \rfloor$ entries of table T. Each entry needs no more than $2+t$ words, where t is the number of items defining the state. If $\overline{p}_{i_1}, \ldots, \overline{p}_{i_t}$ are the scaled profits of such items, we have $t \leq q / \min \{\overline{p}_{i_1}, \ldots, \overline{p}_{i_t}\} \leq 3/\varepsilon$. Hence the overall space complexity of IK(ε) is $O(n)$ (for the input) + $O(1/\varepsilon^3)$.

Theorem 2.4 (Ibarra and Kim, 1975) *For any instance of KP, $(z - z^h)/z \leq \varepsilon$, where z is the optimal solution value and z^h the value returned by* IK(ε).

Proof. If the algorithm terminates in the initial phase with $z^h = \sum_{j=1}^{s-1} p_j$ then z^h gives the optimal solution. Otherwise, let $\{i_1, \ldots, i_k\}$ be the (possibly empty) set of items with $p_{i_l} > \frac{1}{3}\varepsilon \tilde{z}$ in the optimal solution, i.e.

$$z = \sum_{l=1}^{k} p_{i_l} + \alpha,$$

where α is a sum of profits of items in S. Defining $\overline{p}^* = \sum_{l=1}^{k} \overline{p}_{i_l}$ and $w^* = \sum_{l=1}^{k} w_{i_l}$, we have, at the end of the dynamic programming phase, $T(\overline{p}^*) \neq$ "*undefined*" and $W(\overline{p}^*) \leq w^*$ (since $W(i)$ is never increased by the algorithm). Let $L(\overline{p}^*) = \{j_1, \ldots, j_h\}$. (This implies that $\overline{p}^* = \sum_{l=1}^{h} \overline{p}_{j_l}$ and $W(\overline{p}^*) = \sum_{l=1}^{h} w_{j_l}$.) Then the sum $\overline{z} = \sum_{l=1}^{h} p_{j_l} + \beta$, where β is a sum of profits of elements in S, has been considered in the greedy phase (when $i = \overline{p}^*$), so $z^h \geq \overline{z}$. Observe that $\overline{p}_j = \lfloor p_j/\delta \rfloor \geq 3/\varepsilon$, from which $\overline{p}_j\delta \leq p_j \leq (\overline{p}_j + 1)\delta = \overline{p}_j\delta(1 + 1/\overline{p}_j) \leq \overline{p}_j\delta(1 + \varepsilon/3)$. It follows that

$$\overline{p}^*\delta + \alpha \leq z \leq \overline{p}^*\delta(1 + \tfrac{1}{3}\varepsilon) + \alpha,$$

$$\overline{p}^*\delta + \beta \leq \overline{z} \leq \overline{p}^*\delta(1 + \tfrac{1}{3}\varepsilon) + \beta,$$

from which

$$\frac{z - \overline{z}}{z} \leq \frac{\overline{p}^*\delta\varepsilon/3 + (\alpha - \beta)}{z} \leq \tfrac{1}{3}\varepsilon + \frac{\alpha - \beta}{z}.$$

Since $W(\overline{p}^*) \leq w^*$ and the items in S are ordered by decreasing p_j/w_j ratios, it follows that $(\alpha - \beta)$ cannot be greater than the maximum profit of an item in S, i.e. $\alpha - \beta \leq \frac{1}{3}\varepsilon\tilde{z}$. Hence $(z - \overline{z})/z \leq \frac{1}{3}\varepsilon(1 + \tilde{z}/z)$. Since $\overline{z} \leq z^h$ and $\tilde{z} \leq 2z$, then $(z - z^h)/z \leq \varepsilon$. □

Example 2.5 (continued)

We apply IK(ε) with $\varepsilon = \frac{1}{2}$.

$s = 3$;

$\tilde{z} = 1200$;

$\delta = \dfrac{100}{3}$;

$S = \emptyset$ (items with $p_j \leq \varepsilon\tilde{z}/3 = 200$ will be inserted in S);

$T(0) = (\emptyset, 0, 0)$;

$q = 36$;

dynamic programming phase:

$j = 1: \ \overline{p}_1 = 10,\ T(10) = (\{1\}, 350, 25)$;

$j = 2: \ \overline{p}_2 = 12,\ T(22) = (\{1, 2\}, 750, 60)$,

$T(12) = (\{2\}, 400, 35)$;

$j = 3: \ \overline{p}_3 = 13,\ T(25) = (\{2, 3\}, 850, 80)$,

$$T(23) = (\{1, 3\}, 800, 70),$$

$$T(13) = (\{3\}, 450, 45);$$

$j = 4, \ldots, 8: \; S = \{4, 5, 6, 7, 8\};$

greedy phase:

for all the entries of table T save $T(23)$ and $T(25)$, we have $c - W(i) \geq \sum_{j \in S} w_j = 37$. Hence the best solution produced by such states is $P(22) + \sum_{j \in S} p_j = 858$. $T(23)$ gives $P(23) + \sum_{j \in \{4,5,6\}} p_j = 898$; $T(25)$ gives $P(25) + \sum_{j \in \{4,6,7,8\}} p_j = 888$. It follows that $z^h = 898$, $X^h = \{1, 3, 4, 5, 6\}$.

The solution does not change for all values $\varepsilon \geq \frac{1}{50}$. For $\varepsilon < \frac{1}{50}$, we have $\varepsilon \tilde{z}/3 < 8$, so items 1–6 are considered "large" and the algorithm finds the optimal solution using entry $T(i) = (\{1, 3, 4, 5\}, 890, 100)$. The value of q, however, is at least 22 500 instead of 36. □

Ibarra and Kim (1975) have also proposed a modified implementation having improved time complexity $O(n \log n) + O((1/\varepsilon^4) \log(1/\varepsilon))$, with the second term independent of n. Further improvements have been obtained by Lawler (1979), who used a median-finding routine (to eliminate sorting) and a more efficient scaling technique to obtain time complexity $O(n \log(1/\varepsilon) + 1/\varepsilon^4)$ and space complexity $O(n + 1/\varepsilon^3)$. Magazine and Oguz (1981) have further revised the Lawler (1979) scheme, obtaining time complexity $O(n^2 \log(n/\varepsilon))$ and space complexity $O(n/\varepsilon)$.

A fully polynomial-time approximation scheme for the minimization version of KP was found, independently of the Ibarra–Kim result, by Babat (1975). Its time and space complexity of $O(n^4/\varepsilon)$ was improved to $O(n^2/\varepsilon)$ by Gens and Levner (1979).

Note that the core memory requirements of the fully polynomial-time approximation schemes depend on ε and can become impractical for small values of this parameter. On the contrary, the space complexity of Sahni's polynomial-time approximation scheme is $O(n)$, independently of r.

2.8.3 Probabilistic analysis

The first probabilistic result for KP was obtained by d'Atri (1979). Assuming that profits and weights are independently drawn from the uniform distribution over $\{1, 2, \ldots, n\}$, and the capacity from the uniform distribution over $\{1, 2, \ldots, kn\}$ (k an integer constant), he proved that there exists an $O(n)$ time algorithm giving the optimal solution with probability tending to 1 as $n \to \infty$.

Lueker (1982) investigated the properties of the average value of $(z(C(KP)) - z(KP))$ (difference between the solution value of the continuous relaxation and the optimal solution value of KP). Assuming that profits and weights are independently generated from the uniform distribution between 0 and 1 by a Poisson process with n as the expected number of items, and that the capacity is $c = \beta n$ for some constant β, he proved that:

(a) if $\beta > \frac{1}{2}$ then all items fit in the knapsack with probability tending to 1, so the question is trivial;

(b) if $\beta \leq \frac{1}{2}$ then the expected value of $(z(C(KP)) - z(KP))$ is $O(\log^2 n/n)$ and $\Omega(1/n)$.

Goldberg and Marchetti-Spaccamela (1984) improved the $\Omega(1/n)$ lower bound to $\Omega(\log^2 n/n)$, thus proving that the expected value of the difference is $\Theta(\log^2 n/n)$. In addition, they proved that, for every fixed $\varepsilon > 0$, there is a polynomial-time algorithm which finds the optimal solution to KP with probability at least $1 - \varepsilon$. (As a function of $1/\varepsilon$, the running time of the algorithm is exponential.)

Meanti, Rinnooy Kan, Stougie and Vercellis (1989) have determined, for the same probabilistic model, the expected value of the critical ratio p_s/w_s as a function of β, namely $1/\sqrt{6\beta}$ for $0 < \beta < \frac{1}{6}$, $\frac{3}{2} - 3\beta$ for $\frac{1}{6} \leq \beta < \frac{1}{2}$. The result has been used by Marchetti-Spaccamela and Vercellis (1987) to analyse the probabilistic behaviour of an *on-line* version of the greedy algorithm. (An on-line algorithm for KP is required to decide whether or not to include each item in the knapsack as it is input, i.e. as its profit and weight become known.)

The probabilistic properties of different greedy algorithms for KP have been studied in Szkatula and Libura (1987).

2.9 EXACT ALGORITHMS FOR LARGE-SIZE PROBLEMS

As will be shown in Section 2.10, many instances of KP can be solved by branch-and-bound algorithms for very high values of n. For such problems, the preliminary sorting of the items requires, on average, a comparatively high computing time (for example, when $n > 2000$ the sorting time is about 80 per cent of the total time required by the algorithm of Section 2.5.2). In the present section we examine algorithms which do not require preliminary sorting of all the items.

The first algorithm of this kind was presented by Balas and Zemel (1980) and is based on the so-called "core problem". Suppose, without loss of generality, that $p_j/w_j > p_{j+1}/w_{j+1}$ for $j = 1, \dots, n-1$, and, for an optimal solution (x_j^*), define the *core* as

$$C = \{j_1, \dots, j_2\},$$

where

$$j_1 = \min\{j : x_j^* = 0\}, \qquad j_2 = \max\{j : x_j^* = 1\};$$

the *core problem* is then defined as

$$\text{maximize} \quad \tilde{z} = \sum_{j \in C} p_j x_j$$

$$\text{subject to} \quad \sum_{j \in C} w_j x_j \leq c - \sum_{j \in \{i : p_i/w_i > p_{j_1}/w_{j_1}\}} w_j,$$

$$x_j = 0 \text{ or } 1, \qquad \text{for } j \in C.$$

In general, for large problems, the size of the core is a very small fraction of n. Hence, if we knew "a priori" the values of j_1 and j_2, we could easily solve the complete problem by setting $x_j^* = 1$ for all $j \in J1 = \{k : p_k/w_k > p_{j_1}/w_{j_1}\}$, $x_j^* = 0$ for all $j \in J0 = \{k : p_k/w_k < p_{j_2}/w_{j_2}\}$ and solving the core problem through any branch-and-bound algorithm (so that only the items in C would have to be sorted). Notice that $J1$ and $J0$ are conceptually close to the sets of the same name determined by reduction procedures.

Indices j_1 and j_2 cannot be "a priori" identified, but a good approximation of the core problem can be obtained if we consider that, in most cases, given the critical item s, we have $j_1 \geq s - (\vartheta/2)$ and $j_2 \leq s + (\vartheta/2)$ for some $\vartheta \ll n$.

2.9.1 The Balas–Zemel algorithm

Balas and Zemel (1980) proposed the following procedure for determining, given a prefixed value ϑ, an approximate partition $(J1, C, J0)$ of N. The methodology is very close to that used in Section 2.2.2 to determine the critical item s and the continuous solution $(\overline{x}_j)$, so we only give the statements differing from the corresponding ones in procedure CRITICAL_ ITEM:

procedure BZC:
input: $n, c, (p_j), (w_j), \vartheta$;
output: $J1, C, (\overline{x}_j)$;
begin
  ...
  while *partition* = "no" and $|JC| > \vartheta$ **do**
    begin
      determine the median r_t of the first 3 ratios p_j/w_j in JC;
      ...
    end;
  if $|JC| \leq \vartheta$ **then**
    begin
      $C := JC$;
      sort the items in C according to decreasing p_j/w_j ratios;
      determine the critical item s and the solution $(\overline{x}_j)$ of the continuous relaxation through the Dantzig method applied to the items in C with the residual capacity $\overline{c}$
    end
  else
    begin
      let $E = \{e_1, \ldots, e_q\}$;

$\sigma := \min \{j : \sum_{i=1}^{j} w_{e_i} > \overline{c} - c'\}$;
$s := e_\sigma$;
for each $j \in J1 \cup G \cup \{e_1, \dots, e_{\sigma-1}\}$ **do** $\overline{x}_j := 1$;
for each $j \in J0 \cup L \cup \{e_{\sigma+1}, \dots, e_q\}$ **do** $\overline{x}_j := 0$;
$\overline{x}_s := (c - \sum_{j \in \{1,\dots,n\} \setminus \{s\}} w_j \overline{x}_j)/w_s$;
define C as a sorted subset of JC such that $|C| = \vartheta$ and s is contained, if possible, in the middle third of C, and correspondingly enlarge set $J1$
end
end.

Determining the median of the first three ratios (instead of that of all the ratios) in JC increases the time complexity of the algorithm to $O(n^2)$, but is indicated in Balas and Zemel (1980) as the method giving the best experimental results. They had also conjectured that the expected size of the core problem is constant, and experimentally determined it as $\vartheta = 25$. The conjecture has been contradicted by Goldberg and Marchetti-Spaccamela (1984), who proved that the expected core problem size grows (very slowly) with n.

The Balas–Zemel method also makes use of a heuristic procedure H and a reduction procedure R. These can be summarized as follows:

procedure H:
input: $C, J1$;
output: $z, (x_j)$;
begin
given an approximate core problem C and a set $J1$ of items j such that x_j is fixed to 1, find an approximate solution for C by using dominance relations between the items;
define the corresponding approximate solution (x_j), and its value z, for KP
end.

procedure R:
input: C;
output: $J1', J0'$;
begin
fix as many variables of C as possible by applying the reduction test of algorithm FPDHR, then that of algorithm IKR (see Section 2.7), modified so as to compute an upper bound on the continuous solution value when the items are not sorted;
define subsets $J1'$ and $J0'$, containing the variables fixed, respectively, to 1 and to 0
end.

The Balas–Zemel idea is first to solve, without sorting, the continuous relaxation of KP, thus determining the Dantzig upper bound (see Section 2.2.1), and then searching for heuristic solutions of approximate core problems giving the upper bound value for KP. When such attempts fail, the reduced problem is solved through an exact procedure. The algorithm can be outlined as follows (γ is a given threshold value for which Balas and Zemel used $\gamma = 50$).

procedure BZ:
input: $n, c, (p_j), (w_j), \vartheta, \gamma$;
output: $z, (x_j)$;
begin
  call BZC ;
  $z^c := \sum_{j=1}^{n} p_j \bar{x}_j$;
  call H ;
  if $z = \lfloor z^c \rfloor$ **then return**;
  $C := \{1, \dots, n\}$;
  call R;
  $J1 := J1'$;
  $J0 := J0'$;
  $C := C \backslash (J1 \cup J0)$ (**comment**: new core);
  if $|C| > \gamma$ **then**
    begin
      call H ;
      if $z = \lfloor z^c \rfloor$ **then return**;
      call R;
      $J1 := J1 \cup J1'$;
      $J0 := J0 \cup J0'$;
      $C := C \backslash (J1' \cup J0')$ (**comment**: reduced core);
    end;
  sort the items in C according to decreasing p_j / w_j ratios;
  exactly solve the core problem through the Zoltners (1978) algorithm;
  define the corresponding values of z and (x_j) for KP
end.

Two effective algorithms for solving KP without sorting all the items have been derived from the Balas–Zemel idea by Fayard and Plateau (1982) and Martello and Toth (1988).

2.9.2 The Fayard–Plateau algorithm

The algorithm, published together with an effective Fortran implementation (see Fayard and Plateau (1982)), can be briefly described as follows.

procedure FP:
input: $n, c, (p_j), (w_j)$;
output: $z, (x_j)$;
begin
  $N := \{1, \dots, n\}$;

use a procedure similar to CRITICAL_ ITEM (see Section 2.2.2) to determine the critical item s and the subset $J1 \subset N$ such that, in the continuous solution of KP, $x_j = 1$ for $j \in J1$;
$\overline{c} := c - \sum_{j \in J1} w_j$;
$z^c := \sum_{j \in J1} p_j + \overline{c} p_s / w_s$;
apply the greedy algorithm (without sorting) to the items in $N \backslash J1$ with the residual capacity $\overline{c}$, and let (x_j) $(j \in N \backslash J1)$ be the approximate solution found;
$z := \sum_{j \in J1} p_j + \sum_{j \in N \backslash J1} p_j x_j$;
if $z = \lfloor z^c \rfloor$ **then return** ;
apply reduction algorithm FPDHR (see Section 2.7), defining sets $J1'$ and $J0'$;
$C := N \backslash (J1' \cup J0')$ (**comment**: reduced problem);
sort the items in C according to increasing values of $|\tilde{p}_j| = |p_j - w_j p_s / w_s|$;
exactly solve the reduced problem through a specific enumerative technique;
define the corresponding values of z and (x_j) for KP
end.

2.9.3 The Martello–Toth algorithm

The Martello and Toth (1988) algorithm can be sketched as follows.

Step 1. Partition N into $J1, J0$ and C through a modification of the Balas–Zemel method. Sort the items in C.

Step 2. Exactly solve the core problem, thus obtaining an approximate solution for KP, and compute upper bound U_6 (see Section 2.3.3). If its value equals that of the approximate solution then this is clearly optimal: stop. Otherwise

Step 3. Reduce KP with no further sorting: if all variables x_j such that $j \in J1$ or $j \in J0$ are fixed (respectively to 1 and to 0), then we have it that C is the exact core, so the approximate solution of Step 2 is optimal: stop. Otherwise

Step 4. Sort the items corresponding to variables not fixed by reduction and exactly solve the corresponding problem.

The algorithm improves upon the previous works in four main respects:

(a) the approximate solution determined at Step 2 is more precise (often optimal); this is obtained through more careful definition of the approximate core and through exact (instead of heuristic) solution of the corresponding problem;

(b) there is a higher probability that such an approximate solution can be proved to be optimal either at Step 2 (because of a tighter upper bound computation) or at Step 3 (missing in previous works);

(c) the procedures for determining the approximate core (Step 1) and reducing KP (Step 3) have been implemented more efficiently;

(d) the exact solution of the subproblems (Steps 2 and 4) has been obtained by adapting an effective branch-and-bound algorithm (procedure MT1 of Section 2.5.2).

Step 1

The procedure to determine the approximate core problem receives in input four parameters: ϑ (desired core problem size), α, β (tolerances) and η (bound on the number of iterations). It returns a partition $(J1, C, J0)$ of N, where C defines an approximate core problem having residual capacity $\overline{c} = c - \sum_{j \in J1} w_j$, such that

(i) $(1 - \alpha)\vartheta \leq |C| \leq (1 + \beta)\vartheta$,

(ii) $\sum_{j \in C} w_j > \overline{c} > 0$,

(iii) $\max \{p_k/w_k : k \in J0\} \leq p_j/w_j \leq \min \{p_k/w_k : k \in J1\}$ for all $j \in C$.

$J1$ and $J0$ are initialized to empty, and C to N. At any iteration we try to move elements from N to $J1$ or $J0$, until $|C|$ is inside the prefixed range. Following Balas and Zemel (1980), this is obtained by partitioning (through a tentative value λ) set C into three sets of items j such that p_j/w_j is less than λ (set L), equal to λ (set E) or greater than λ (set G). Three possibilities are then considered, according to the value of the current residual capacity $\overline{c}$:

(a) $\sum_{j \in G} w_j \leq \overline{c} < \sum_{j \in G \cup E} w_j$, i.e., $\lambda = p_s/w_s$: if $|E|$ is large enough, the desired core is defined; otherwise λ is increased or decreased, according to the values of $|G|$ and $|L|$, so that $|C|$ results closer to the desired size at the next iteration;

(b) $\sum_{j \in G} w_j > \overline{c}$, i.e., $\lambda < p_s/w_s$: if $|G|$ is large enough we move the elements of $L \cup E$ from C to $J0$ and increase λ; otherwise we decrease λ so that, at the next iteration, $|G|$ results larger;

(c) $\sum_{j \in G \cup E} w_j < \overline{c}$, i.e., $\lambda > p_s/w_s$: if $|L|$ is large enough we move the elements of $G \cup E$ from C to $J1$ and decrease λ; otherwise we increase λ so that, at the next iteration, $|L|$ results larger.

In the following description of procedure CORE, $M3(S)$ denotes the median of the profit/weight ratios of the first, last and middle element of S. If the desired C is not obtained within η iterations, execution is halted and the current partition $(J1, C, J0)$ is returned. In this case, however, condition (i) above is not satisfied, i.e. $|C|$ is not inside the prefixed range.

```
procedure CORE:
input: n, c, (p_j), (w_j), ϑ, α, β, η;
output: J1, C, J0;
begin
    J1 := ∅;
    J0 := ∅;
    C := {1, ..., n};
    c̄ := c;
    k := 0;
    λ := M3(C);
    while |C| > (1 + β)ϑ and k < η do
        begin
            G := {j ∈ C : p_j/w_j > λ};
            L := {j ∈ C : p_j/w_j < λ};
            E := {j ∈ C : p_j/w_j = λ};
            c' := Σ_{j∈G} w_j;
            c'' := c' + Σ_{j∈E} w_j;
            if c' ≤ c̄ < c'' then
                if |E| ≥ (1 − α)ϑ then
                    begin
                        let E = {e_1, ..., e_q};
                        σ := min {j : Σ_{i=1}^{j} w_{e_i} > c̄ − c'};
                        s := e_σ;
                        C := {e_r, ..., e_t} with r, t such that
                            t − r + 1 is as close as possible to ϑ
                            and (t + r)/2 to s;
                        J0 := J0 ∪ L ∪ {e_{t+1}, ..., e_q};
                        J1 := J1 ∪ G ∪ {e_1, ..., e_{r−1}}
                    end
                else
                    if |G ∪ E| < ϑ then λ := M3(L)
                    else λ := M3(G)
            else
                if c' > c̄ then
                    if |G| < (1 − α)ϑ then λ := M3(L)
                    else
                        begin
                            J0 := J0 ∪ L ∪ E;
                            C := G;
                            λ := M3(C)
                        end
                else
                    if |L| < (1 − α)ϑ then λ := M3(G)
                    else
                        begin
                            J1 := J1 ∪ G ∪ E;
                            C := L;
                            c̄ := c̄ − c'';
```

```
                    λ := M3(C)
                end;
            k := k + 1
        end
end.
```

The heaviest computations in the "while" loop (partitioning of C and definition of c' and c'') require $O(n)$ time. Hence, if η is a prefixed constant, the procedure runs in linear time.

Steps 2, 4

Exact solutions $(\hat{x}_j)$ of the core problem and of the reduced problem are obtained through procedure MT1 of Section 2.5.2, modified so as also to compute, if required, the value u of upper bound U_6 (Section 2.3.3) for KP. We refer to this procedure as MT1′ and call it by giving the sets C (free items) and $J1$ (items j such that x_j is fixed to 1).

procedure MT1′:
input: $n, c, (p_j), (w_j), C, J1$, *bound*;
output: $(\hat{x}_j), u$;
begin
  define the sub-instance KP′ consisting of the items in C with residual capacity $c - \sum_{j \in J1} w_j$;
  if *bound* = "no" **then call** MT1 for KP′
  else call MT1 for KP′ with determination of $u = U_6$;
  let $(\hat{x}_j)$ be the solution vector returned by MT1
end.

Step 3

Reduction without sorting is obtained through the following procedure, which receives in input the partition determined at Step 1 (with only the items in C sorted according to decreasing p_j/w_j ratios) and the value z^h of the approximate solution found at Step 2. The procedure defines sets $\overline{J1}$ and $\overline{J0}$ according to the same rules as in procedure MTR (Section 2.7), but computing weaker bounds u_j^0 and u_j^1 when the current critical item $\overline{s}$ is not in C.

procedure MTR′:
input: $n, c, (p_j), (w_j), z^h, J1, C, J0$;
output: $\overline{J1}, \overline{J0}$;
begin
  comment: it is assumed that the items in C are $1, 2, \ldots, f$ ($f = |C|$), sorted according to decreasing p_j/w_j ratios;
  $\overline{c} := c - \sum_{j \in J1} w_j$;
  $\overline{p} := \sum_{j \in J1} p_j$;

```
for j := 1 to f do compute w̄_j = Σ_{i=1}^{j} w_i and p̄_j = Σ_{i=1}^{j} p_i;
find, through binary search, s ∈ C such that w̄_{s-1} ≤ c̄ < w̄_s;
for each j ∈ J1 ∪ {1,...,s} do
    if c̄ + w_j < w̄_f then
        begin
            find, through binary search, s̄ ∈ C such that
              w̄_{s̄-1} ≤ c̄ + w_j < w̄_s̄;
            c̿ := c̄ + w_j − w̄_{s̄-1};
            u_j^0 := p̄ − p_j + p̄_{s̄-1} + max (⌊c̿ p_{s̄+1}/w_{s̄+1}⌋,
                                  ⌊p_s̄ − (w_s̄ − c̿) p_{s̄-1}/w_{s̄-1}⌋);
            z^h := max (z^h, p̄ − p_j + p̄_{s̄-1})
        end
    else
        begin
            u_j^0 := p̄ − p_j + p̄_f + ⌊(c̄ + w_j − w̄_f) p_f/w_f⌋;
            z^h := max (z^h, p̄ − p_j + p̄_f)
        end;
for each j ∈ J0 ∪ {s,...,f} do
    if c̄ − w_j ≥ w̄_1 then
        begin
            find, through binary search, s̄ ∈ C such that
              w̄_{s̄-1} ≤ c̄ − w_j < w̄_s̄;
            c̿ := c̄ − w_j − w̄_{s̄-1};
            u_j^1 := p̄ + p_j + p̄_{s̄-1} + max (⌊c̿ p_{s̄+1}/w_{s̄+1}⌋,
                                  ⌊p_s̄ − (w_s̄ − c̿) p_{s̄-1}/w_{s̄-1}⌋);
            z^h := max (z^h, p̄ + p_j + p̄_{s̄-1})
        end
    else
        begin
            u_j^1 := ⌊p̄ + p_j + (c̄ − w_j) p_1/w_1⌋;
            if c̄ − w_j ≥ 0 then z^h := max (z^h, p̄ + p_j)
        end;
J̄0 := {j ∈ J0 ∪ {s,...,f} : u_j^1 ≤ z^h};
J̄1 := {j ∈ J1 ∪ {1,...,s} : u_j^0 ≤ z^h}
end.
```

The heaviest computations are involved in the two "for each" loops: for $O(n)$ times a binary search, of time complexity $O(\log|C|)$, is performed. The overall time complexity is thus $O(n\log|C|)$, i.e. $O(n)$ for fixed $|C|$.

Algorithm

The algorithm exactly solves KP through Steps 1–4, unless the size of core C determined at Step 1 is too large. If this is the case, the solution is obtained through standard sorting, reduction and branch-and-bound. On input, the items are not assumed to be sorted.

```
procedure MT2:
input: n, c, (p_j), (w_j), ϑ, α, β, η;
output: z, (x_j);
begin
    for j := 1 to n do x_j := 0;
    comment: Step 1;
    call CORE;
    if |C| ≤ (1 − α)n then
        begin
            sort the items in C by decreasing p_j/w_j ratios;
            comment: Step 2;
            bound := "yes";
            call MT1';
            z^h := Σ_{j∈J1} p_j + Σ_{j∈C} p_j x̂_j;
            if z^h = u then
                for each j ∈ J1 ∪ {k ∈ C : x̂_k = 1} do x_j := 1
            else (comment: Step 3)
                begin
                    call MTR';
                    if J̄1 ⊇ J1 and J̄0 ⊇ J0 then
                        for each j ∈ J1 ∪ {k ∈ C : x̂_k = 1} do x_j := 1
                    else (comment: Step 4)
                        begin
                            C := {1, ..., n}\(J̄1 ∪ J̄0);
                            sort the items in C according to
                                decreasing p_j/w_j ratios;
                            bound := "no";
                            J1 := J̄1;
                            call MT1' ;
                            for each j ∈ J̄1 ∪ {k ∈ C : x̂_k = 1} do x_j := 1
                        end
                end
        end
    else (comment: standard solution)
        begin
            sort all the items according to decreasing p_j/w_j ratios;
            call MTR;
            z^h := l;
            C := {1, ..., n}\(J1 ∪ J0);
            bound:= "no";
            call MT1';
            for each j ∈ J1 ∪ {k ∈ C : x̂_k = 1} do x_j := 1
        end;
```

```
    z := ∑_{j=1}^n p_j x_j;
    if z < z^h then
        begin
            define the solution vector (x_j) corresponding to z^h;
            z := z^h
        end
end.
```

On the basis of the computational experiments reported in the next section, the four parameters needed by MT2 have been determined as

$$\vartheta = \begin{cases} n & \text{if } n < 200, \\ 2\sqrt{n} & \text{otherwise;} \end{cases}$$

$$\alpha = 0.2;$$

$$\beta = 1.0;$$

$$\eta = 20.$$

The Fortran implementation of MT2 is included in the present volume.

2.10 COMPUTATIONAL EXPERIMENTS

In this section we analyse the experimental behaviour of exact and approximate algorithms for KP on sets of randomly generated test problems. Since the difficulty of such problems is greatly affected by the correlation between profits and weights, we consider three randomly generated data sets:

uncorrelated: p_j and w_j uniformly random in $[1, v]$;

weakly correlated: w_j uniformly random in $[1, v]$,
p_j uniformly random in $[w_j - r, w_j + r]$;

strongly correlated: w_j uniformly random in $[1, v]$,
$p_j = w_j + r$.

Increasing correlation means decreasing value of the difference $\max_j\{p_j/w_j\} - \min_j\{p_j/w_j\}$, hence increasing expected difficulty of the corresponding problems. According to our experience, weakly correlated problems are closer to real world situations.

For each data set we consider two values of the capacity: $c = 2v$ and $c = 0.5\sum_{j=1}^n w_j$. In the first case the optimal solution contains very few items, so the generated instances are expected to be easier than in the second case, where about half of the items are in the optimal solution. (Further increasing the value of c does not significantly increase the computing times.)

2.10.1 Exact algorithms

We give separate tables for small-size problems ($n \leq 200$) and large-size problems ($n \geq 500$).

We compare the Fortran IV implementations of the following algorithms:

HS = Horowitz and Sahni (1974), Section 2.5.1;

MTR+HS = HS preceded by reduction procedure MTR of Section 2.7;

NA = Nauss (1976), with its own reduction procedure;

MT1 = Martello and Toth (1977a), Section 2.5.2;

MTR+MT1 = Martello and Toth (1977a) preceded by MTR;

MTR+DPT = Toth (1980), Section 2.6.3, preceded by MTR;

BZ = Balas and Zemel (1980), Section 2.9.1, with its own reduction procedure;

FP = Fayard and Plateau (1982), Section 2.9.2, with its own reduction procedure;

MT2 = Martello and Toth (1988), Section 2.9.3, with MTR and MTR′.

NA, MT1, FP and MT2 are published codes, whose characteristics are given in Table 2.1. HS, MTR and DPT have been coded by us. For BZ we give the computing times presented by the authors.

Table 2.1 Fortran codes for KP

Authors	Core memory	Number of statements	List
Nauss (1976)	$8n$	280	Available from the author
Martello and Toth (1977a)	$8n$	280	This volume (also in Martello and Toth (1978))
Fayard and Plateau (1982)	$7n$	600	In Fayard and Plateau (1982)
Martello and Toth (1988)	$8n$	1400	This volume

All runs (except those of Table 2.8) were executed on a CDC-Cyber 730. For each data set, value of c and value of n, the tables give the average running time, expressed in seconds, computed over 20 problem instances. Since Balas and Zemel (1980) give times obtained on a CDC-6600, which we verified to be at least two times faster than the CDC-Cyber 730 on problems of this kind, the times given in the tables for BZ are those reported by the authors multiplied by 2.

Code FP includes its own sorting procedure. The sortings needed by HS, NA, MT1, DPT and MT2 were obtained through a subroutine (included in MT2), derived

Table 2.2 Sorting times. CDC-Cyber 730 in seconds. Average times over 20 problems

n	50	100	200	500	1000	2000	5000	10 000
time	0.008	0.018	0.041	0.114	0.250	0.529	1.416	3.010

Table 2.3 Uncorrelated problems: p_j and w_j uniformly random in [1,100]. CDC-Cyber 730 in seconds. Average times over 20 problems

c	n	HS	MTR +HS	NA	MT1	MTR +MT1	FP	MTR +DPT
	50	0.022	0.013	0.015	0.015	0.012	0.013	0.013
200	100	0.039	0.024	0.025	0.026	0.025	0.018	0.029
	200	0.081	0.050	0.055	0.051	0.050	0.032	0.055
	50	0.031	0.016	0.015	0.016	0.013	0.013	0.020
$0.5\sum_{j=1}^{n} w_j$	100	0.075	0.028	0.029	0.030	0.026	0.021	0.043
	200	0.237	0.065	0.073	0.068	0.057	0.053	0.090

Table 2.4 Weakly correlated problems: w_j uniformly random in [1,100], p_j in $[w_j-10, w_j+10]$. CDC-Cyber 730 in seconds. Average times over 20 problems

c	n	HS	MTR +HS	NA	MT1	MTR +MT1	FP	MTR +DPT
	50	0.031	0.018	0.019	0.017	0.014	0.016	0.022
200	100	0.049	0.029	0.038	0.032	0.024	0.023	0.041
	200	0.091	0.052	0.060	0.055	0.048	0.030	0.066
	50	0.038	0.025	0.035	0.022	0.020	0.021	0.071
$0.5\sum_{j=1}^{n} w_j$	100	0.079	0.042	0.086	0.040	0.031	0.039	0.158
	200	0.185	0.070	0.151	0.069	0.055	0.057	0.223

Table 2.5 Strongly correlated problems: w_j uniformly random in [1,100], $p_j = w_j + 10$. CDC-Cyber 730 in seconds. Average times over 20 problems

c	n	HS	MTR +HS	NA	MT1	MTR +MT1	FP	MTR +DPT
	50	0.165	0.101	0.117	0.028	0.025	0.047	0.041
200	100	1.035	0.392	0.259	0.052	0.047	0.096	0.070
	200	3.584	2.785	3.595	0.367	0.311	0.928	0.111
	50	time	time	time	4.870	4.019	17.895	0.370
$0.5\sum_{j=1}^{n} w_j$	100	—	—	—	time	time	time	1.409
	200	—	—	—	—	—	—	3.936

from subroutine SORTZV of the CERN Library, whose experimental behaviour is given in Table 2.2. All the times in the following tables include sorting and reduction times.

Tables 2.3, 2.4 and 2.5 compare algorithms HS, MTR+HS, NA, MT1, MTR+MT1, FP and MTR+DPT on small-size problems (we do not give the times of MT2, which are almost equal to those of MTR+MT1). For all data sets, $v = 100$ and $r = 10$. Table 2.3 refers to uncorrelated problems, Table 2.4 to weakly correlated problems. All algorithms solved the problems very quickly with the exception of HS and, for weakly correlated problems, MTR+DPT. MT1 is only slightly improved by previous application of MTR, contrary to what happens for HS. Table 2.5 refers to strongly correlated problems. Because of the high times generally involved, a time limit of 500 seconds was assigned to each algorithm for solution of the 60 problems generated for each value of c. The dynamic programming approach appears clearly superior to all branch-and-bound algorithms (among which MT1 has the best performance).

For large-size instances we do not consider strongly correlated problems, because of the impractical times involved. Tables 2.6 and 2.7 compare algorithms MT1, BZ, FP and MT2. Dynamic programming is not considered because of excessive memory requirements, HS and NA because of clear inferiority. The problems were generated with $v = 1\,000$, $r = 100$ and $c = 0.5 \sum_{j=1}^{n} w_j$.

FP is fast for $n \leq 2\,000$ but very slow for $n \geq 5\,000$, while BZ has the opposite behaviour. MT2 has about the same times as FP for $n \leq 2\,000$, the same times as BZ for $n = 5\,000$, and slightly higher than BZ for $n = 10\,000$, so it can be considered, on average, the best code. MT1, which is not designed for large

Table 2.6 Uncorrelated problems: p_j and w_j uniformly random in [1,1000]; $c = 0.5 \sum_{j=1}^{n} w_j$. CDC-Cyber 730 in seconds. Average times over 20 problems

n	MT1	BZ	FP	MT2
500	0.199	—	0.104	0.157
1 000	0.381	0.372	0.188	0.258
2 000	0.787	0.606	0.358	0.462
5 000	1.993	0.958	1.745	0.982
10 000	4.265	1.514	7.661	1.979

Table 2.7 Weakly correlated problems: w_j uniformly random in [1,1000], p_j in $[w_j - 100, w_j + 100]$; $c = 0.5 \sum_{j=1}^{n} w_j$. CDC-Cyber 730 in seconds. Average times over 20 problems

n	MT1	BZ	FP	MT2
500	0.367	—	0.185	0.209
1 000	0.663	0.588	0.271	0.293
2 000	1.080	0.586	0.404	0.491
5 000	2.188	0.744	1.782	0.771
10 000	3.856	1.018	19.481	1.608

Table 2.8 Algorithm MT2. w_j uniformly random in [1,1000]; $c = 0.5 \sum_{j=1}^{n} w_j$. HP 9000/840 in seconds. Average times over 20 problems

n	Uncorrelated problems: p_j uniformly random in [1,1000]	Weakly correlated problems: p_j uniformly random in $[w_j - 100, w_j + 100]$
50	0.008	0.015
100	0.016	0.038
200	0.025	0.070
500	0.067	0.076
1 000	0.122	0.160
2 000	0.220	0.260
5 000	0.515	0.414
10 000	0.872	0.739
20 000	1.507	1.330
30 000	2.222	3.474
40 000	2.835	2.664
50 000	3.562	3.492
60 000	4.185	504.935
70 000	4.731	4.644
80 000	5.176	5.515
90 000	5.723	6.108
100 000	7.001	7.046
150 000	9.739	time limit
200 000	14.372	—
250 000	17.135	—

problems, is generally the worst algorithm. However, about 80 per cent of its time is spent in sorting, so its use can be convenient when several problems are to be solved for the same item set and different values of c. A situation of this kind arises for multiple knapsack problems, as will be seen in Section 6.4.

$n = 10\,000$ is the highest value obtainable with the CDC-Cyber 730 computer available at the University of Bologna, because of a core memory limitation of 100 Kwords. Hence, we experimented the computational behaviour of MT2 for higher values of n on an HP 9000/840 with 10 Mbytes available. We used the Fortran compiler with option "-o", producing an object with no special optimization. The results obtained for uncorrelated and weakly correlated problems are shown in Table 2.8. Uncorrelated problems were solved up to $n = 250\,000$ with very regular average times, growing less than linearly with n. Weakly correlated problems show an almost linear growing rate, but less regularity; for high values of n, certain instances required extremely high times (for $n = 60\,000$ one of the instances took almost 3 hours CPU time, for $n = 150\,000$ execution was halted after 4 hours).

2.10.2 Approximate algorithms

In Tables 2.9–2.11 we experimentally compare the polynomial-time approximation scheme of Sahni (Section 2.8.1) and a heuristic version of algorithm MT2

Table 2.9 Uncorrelated problems: p_j and w_j uniformly random in [1,1000]; $c = 0.5 \sum_{j=1}^{n} w_j$. HP 9000/840 in seconds. Average times (average percentage errors) over 20 problems

n	MT2 approx. time (% error)	S(0) time (% error)	S(1) time (% error)	S(2) time (% error)
50	0.004(0.10569)	0.005(5.36560)	0.017(5.13968)	0.319(5.05006)
100	0.009(0.05345)	0.009(2.25800)	0.060(2.21412)	2.454(2.19447)
200	0.015(0.03294)	0.017(1.15739)	0.210(1.12217)	19.376(1.11691)
500	0.029(0.00767)	0.049(0.49120)	1.242(0.47978)	299.593(0.47577)
1 000	0.058(0.00418)	0.105(0.21213)	4.894(0.20748)	—
2 000	0.117(0.00251)	0.224(0.10531)	19.545(0.10338)	—
5 000	0.296(0.00182)	0.618(0.05540)	125.510(0.05488)	—
10 000	0.641(0.00076)	1.320(0.02045)	—	—
20 000	1.248(0.00032)	2.852(0.00897)	—	—
30 000	1.873(0.00016)	4.363(0.00786)	—	—
40 000	2.696(0.00016)	6.472(0.00521)	—	—
50 000	3.399(0.00011)	8.071(0.00428)	—	—
60 000	3.993(0.00009)	9.778(0.00403)	—	—
70 000	4.652(0.00003)	11.420(0.00301)	—	—
80 000	5.307(0.00008)	13.075(0.00329)	—	—
90 000	5.842(0.00016)	14.658(0.00247)	—	—
100 000	6.865(0.00007)	16.347(0.00231)	—	—
150 000	9.592(0.00005)	25.357(0.00156)	—	—
200 000	13.223(0.00008)	35.050(0.00144)	—	—
250 000	16.688(0.00010)	44.725(0.00094)	—	—

(Section 2.9.3). The fully polynomial-time approximation schemes are not included since a limited series of experiments showed a dramatic inferiority of these algorithms (see also Section 4.4.2, where this trend is confirmed for the subset-sum problem).

The heuristic version of MT2 was obtained by halting execution at the end of Step 2, and returning the approximate solution of value z^h. In order to obtain a small core problem, procedure CORE was executed with parameters

$$\vartheta = 5;$$

$$\alpha = 0.0;$$

$$\beta = 1.0;$$

$$\eta = 200.$$

As for the Sahni scheme S(k), we experimented S(0), S(1) and S(2), since the time complexity $O(n^{k+1})$ makes the algorithm impractical for $k \geq 3$.

Tables 2.9, 2.10 and 2.11 give the results for the three data sets, with v = 1 000, r = 100 and $c = 0.5 \sum_{j=1}^{n} w_j$. For each approximate algorithm, we give (in brackets) the average percentage error. This was computed as $100(z - z^a)/z$, where z^a is the approximate solution value and z either the optimal solution value (when

Table 2.10 Weakly correlated problems: w_j uniformly random in [1,1000], p_j in $[w_j - 100, w_j + 100]$; $c = 0.5 \sum_{j=1}^{n} w_j$. HP 9000/840 in seconds. Average times (average percentage errors) over 20 problems

n	MT2 approx. time (% error)	S(0) time (% error)	S(1) time (% error)	S(2) time (% error)
50	0.006(0.17208)	0.004(2.13512)	0.017(1.81004)	0.302(1.77572)
100	0.008(0.04296)	0.008(0.87730)	0.055(0.78573)	2.281(0.76862)
200	0.013(0.06922)	0.015(0.31819)	0.194(0.28838)	17.779(0.28216)
500	0.033(0.01174)	0.046(0.14959)	1.139(0.14300)	273.118(0.14135)
1 000	0.058(0.00774)	0.103(0.08226)	4.432(0.07842)	—
2 000	0.114(0.00589)	0.222(0.03740)	17.626(0.03634)	—
5 000	0.312(0.00407)	0.619(0.01445)	113.527(0.01413)	—
10 000	0.645(0.00261)	1.324(0.00630)	—	—
20 000	1.297(0.00155)	2.802(0.00312)	—	—
30 000	1.943(0.00104)	4.372(0.00216)	—	—
40 000	2.667(0.00052)	6.432(0.00177)	—	—
50 000	3.374(0.00036)	8.013(0.00139)	—	—
60 000	4.544(0.00028)	9.377(0.00095)	—	—
70 000	4.662(0.00040)	11.069(0.00083)	—	—
80 000	6.029(0.00031)	13.041(0.00070)	—	—
90 000	6.249(0.00040)	15.662(0.00071)	—	—
100 000	6.618(0.00017)	16.358(0.00050)	—	—
150 000	10.231(0.00019)	25.530(0.00041)	—	—
200 000	12.991(0.00004)	35.230(0.00027)	—	—
250 000	16.062(0.00009)	45.234(0.00020)	—	—

Table 2.11 Strongly correlated problems: w_j uniformly random in [1,1000], $p_j = w_j + 100$; $c = 0.5 \sum_{j=1}^{n} w_j$. HP 9000/840 in seconds. Average times (average percentage errors) over 20 problems

n	MT2 approx. time (% error)	S(0) time (% error)	S(1) time (% error)	S(2) time (% error)
50	0.008(1.50585)	0.003(3.25234)	0.019(1.68977)	0.340(0.74661)
100	0.008(0.81601)	0.007(1.43595)	0.061(0.73186)	2.574(0.39229)
200	0.015(0.51026)	0.017(0.77478)	0.226(0.40653)	20.877(0.26096)
500	0.029(0.27305)	0.046(0.33453)	1.372(0.17836)	316.804(0.09783)
1 000	0.059(0.10765)	0.111(0.15991)	5.388(0.08409)	—
2 000	0.119(0.06850)	0.236(0.08866)	21.173(0.05196)	—
5 000	0.315(0.02148)	0.614(0.02740)	132.973(0.01421)	—
10 000	0.679(0.01384)	1.341(0.01573)	—	—
20 000	1.266(0.00559)	2.787(0.00694)	—	—
30 000	1.879(0.00512)	4.333(0.00504)	—	—
40 000	2.603(0.00292)	6.022(0.00372)	—	—
50 000	3.182(0.00240)	7.598(0.00239)	—	—
60 000	3.795(0.00224)	9.194(0.00252)	—	—
70 000	4.529(0.00167)	10.760(0.00214)	—	—
80 000	5.090(0.00154)	12.324(0.00185)	—	—
90 000	5.595(0.00115)	13.968(0.00179)	—	—
100 000	6.320(0.00132)	15.569(0.00165)	—	—
150 000	9.141(0.00083)	24.583(0.00082)	—	—
200 000	12.005(0.00077)	34.400(0.00083)	—	—
250 000	15.950(0.00055)	44.001(0.00044)	—	—

available) or an upper bound determined by the approximate version of MT2. The execution of each approximate algorithm was halted as soon as the average computing time exceeded 100 seconds.

Table 2.9 shows that it is not convenient to heuristically solve uncorrelated problems, since the exact version of MT2 requires about the same times as its approximate version, which in turn dominates S(k). The same consideration holds for weakly correlated problems with $n \leq 50\,000$ (Table 2.10); for $n > 50\,000$, the approximate version of MT2 dominates S(0), while S(1) and S(2) have impractical time requirements. Table 2.11 shows that the approximate version of MT2; which dominates S(0), must be recommended for large-size strongly correlated problems; for small values of n, S(1) and S(2) can produce better approximations but require dramatically higher computing times.

The Fortran code corresponding to MT2, included in the volume, allows use either of the exact or the approximate version through an input parameter.

2.11 FACETS OF THE KNAPSACK POLYTOPE

In this section we give an outline of the main results obtained in the study of the knapsack polytope. Since such results did not lead, up to now, to the design of effective algorithms for KP, the purpose of the present section is only to introduce the reader to the principal polyhedral concepts and to indicate the relevant literature concerning knapsack problems. Detailed introductions to the theory of polyhedra can be found in Bachem and Grötschel (1982), Pulleyblank (1983), Schrijver (1986) and Nemhauser and Wolsey (1988), among others.

We start with some basic definitions. Given a vector $a \in \mathbb{R}^n$ and a scalar $a_0 \in \mathbb{R}$, the set $\{x \in \mathbb{R}^n : \sum_{j=1}^n a_j x_j = a_0\}$ is called a *hyperplane*. A hyperplane defines two *halfspaces*, namely $\{x \in \mathbb{R}^n : \sum_{j=1}^n a_j x_j \leq a_0\}$ and $\{x \in \mathbb{R}^n : \sum_{j=1}^n a_j x_j \geq a_0\}$. The intersection of finitely many halfspaces, when it is bounded and non-empty, is called a *polytope*. Hence, polytopes can be written as $P = \{x \in \mathbb{R}^n : \sum_{j=1}^n a_{ij} x_j \leq a_{i0} \quad \text{for } i = 1, \dots, r\}$; alternatively, they can be described as the convex hull of finitely many points, i.e. $P = \text{conv}\,(S)$, with $S \subset \mathbb{R}^n$ and $|S|$ finite. m points $x^1, \dots, x^m \in \mathbb{R}^n$ are called *affinely independent* if the equations $\sum_{k=1}^m \lambda_k x^k = 0$ and $\sum_{k=1}^m \lambda_k = 0$ imply $\lambda_k = 0$ for $k = 1, \dots, m$. The *dimension* of a polytope $P \subset \mathbb{R}^n$, $\dim\,(P)$, is $|\overline{P}| - 1$, where $\overline{P}$ is the largest subset of affinely independent points of P. A subset F of a polytope $P \subset \mathbb{R}^n$ is called a *face* of P if there exists an inequality $\sum_{j=1}^n a_j x_j \leq a_0$ which is satisfied by any $x \in P$ and such that $F = \{x \in P : \sum_{j=1}^n a_j x_j = a_0\}$. In other words, a face is the intersection of the polytope and a hyperplane defining a halfspace containing the polytope itself. A face F of P such that $\dim\,(F) = \dim\,(P) - 1$ is called a *facet* of P. Hence an inequality $\sum_{j=1}^n a_j x_j \leq a_0$ defines a facet of P if (a) it is satisfied by any $x \in P$, and (b) it is satisfied with equality by exactly $\dim\,(P)$ affinely independent $x \in P$. The set of inequalities defining all the distinct facets of a polytope P

constitutes the minimal inequality representation of P. Hence the importance of facets in order to apply linear programming techniques to combinatorial problems.

Coming to KP, its constraint set (conditions (2.2), (2.3)) defines the *knapsack polytope*

$$K = \text{conv}\left\{x \in \mathbb{R}^n : \sum_{j=1}^{n} w_j x_j \leq c, \quad x_j \in \{0, 1\} \quad \text{for } j = 1, \dots, n\right\}.$$

It is easy to verify that, with assumption (2.6) ($w_j \leq c$ for all j),

$$\dim(K) = n.$$

In fact (a) $\dim(K) \leq n$ (obvious), and (b) $\dim(K) \geq n$, since K contains the $n+1$ affinely independent points x^k ($k = 0, \dots, n$), where $x^0 = (0, \dots, 0)$ and x^k corresponds to unit vector e_k ($k = 1, \dots, n$). The two main classes of facets of K are based on *minimal covers* and *(1, k)-configurations*.

A set $S \subset N = \{1, \dots, n\}$ is called a *cover* for K if

$$\sum_{j \in S} w_j > c.$$

A cover is called *minimal* if

$$\sum_{j \in S \setminus \{i\}} w_j \leq c \quad \text{for any} \quad i \in S.$$

The set $E(S) = S \cup S'$, where

$$S' = \{j \in N \setminus S \ : \ w_j \geq \max_{i \in S} \{w_i\}\},$$

is called the *extension* of S to N. Let $\mathcal{S}$ be the family of all minimal covers S for K. Balas and Jeroslow (1972) have shown that constraints (2.2), (2.3) are equivalent to the set of *canonical inequalities*

$$\sum_{j \in E(S)} x_j \leq |S| - 1 \quad \text{for all} \quad S \in \mathcal{S}, \tag{2.46}$$

in the sense that $x \in \{0, 1\}^n$ satisfies (2.2), (2.3) if and only if it satisfies (2.46). Balas (1975), Hammer, Johnson and Peled (1975) and Wolsey (1975) have given necessary and sufficient conditions for a canonical inequality to be a facet of K.

A rich family of facets of K can be obtained by "lifting" facets of lower dimensional polytopes. Given a minimal cover S for K, let $K_S \subset \mathbb{R}^{|S|}$ denote the $|S|$-dimensional polytope

$$K_S = \text{conv}\left\{ x \in \{0,1\}^{|S|} : \sum_{j \in S} w_j x_j \le c \right\}, \tag{2.47}$$

i.e. the subset of K containing only points x such that $x_j = 0$ for all $j \in N \backslash S$. It is known (see, for instance, Balas (1975), Padberg (1975), Wolsey (1975)) that the inequality

$$\sum_{j \in S} x_j \le |S| - 1$$

defines a facet of the lower dimensional polytope K_S. Nemhauser and Trotter (1974) and Padberg (1975) have given a *sequential lifting* procedure to determine integer coefficients β_j $(j \in N \backslash S)$ such that the inequality

$$\sum_{j \in S} x_j + \sum_{j \in N \backslash S} \beta_j x_j \le |S| - 1$$

defines a facet of K. Calculating these coefficients requires solution of a sequence of $|N \backslash S|$ 0-1 knapsack problems. Furthermore, the facet obtained depends on the sequence in which indices $j \in N \backslash S$ are considered. Zemel (1978) and Balas and Zemel (1978) have given a characterization of the entire class of facets associated with minimal covers, and a *simultaneous lifting* procedure to obtain them. These facets have in general fractional coefficients (those with integer coefficients coincide with the facets produced by sequential lifting).

A richer class of facetial inequalities of K is given by $(1,k)$-configurations (Padberg, 1979, 1980). Given a subset $M \subset N$ and $t \in N \backslash M$, define the set $S = M \cup \{t\}$. S is a *$(1,k)$-configuration* for K if (a) $\sum_{j \in M} w_j \le c$ and (b) $Q \cup \{t\}$ is a minimal cover for every $Q \subseteq M$ with $|Q| = k$, where k is any given integer satisfying $2 \le k \le |M|$. Note that if $k = |M|$, a $(1,k)$-configuration is a minimal cover for K (and, conversely, any minimal cover S can be expressed as a $(1,k)$-configuration, with $k = |S| - 1$, for any $t \in S$). Padberg (1980) proved that, given a $(1,k)$-configuration $S = M \cup \{t\}$ of K, the complete and irredundant set of facets of the lower dimensional polytope K_S (see 2.47) is given by the inequalities

$$(r - k + 1)x_t + \sum_{j \in S(r)} x_j \le r,$$

where $S(r) \subseteq M$ is any subset of cardinality r, and r is any integer satisfying $k \le r \le |M|$. Sequential or simultaneous lifting procedures can then be used to obtain facets of the knapsack polytope K.

Recently, Gottlieb and Rao (1988) have studied a class of facets of K, containing fractional coefficients, which can be derived from disjoint and overlapping minimal covers and $(1,k)$-configurations. For such class, they have given necessary and sufficient conditions which can easily be verified without use of the computationally heavy simultaneous lifting procedures. The computational complexity of lifted inequalities has been analysed by Hartvigsen and Zemel (1987) and Zemel (1988).

2.12 THE MULTIPLE-CHOICE KNAPSACK PROBLEM

The *Multiple-Choice Knapsack Problem* (MCKP), also known as the *Knapsack Problem with Generalized Upper Bound (GUB) Constraints*, is a 0-1 knapsack problem in which a partition $N_1, \dots, N_r$ of the item set N is given, and it is required that exactly one item per subset is selected. Formally,

$$\text{maximize} \quad z = \sum_{j=1}^{n} p_j x_j \tag{2.48}$$

$$\text{subject to} \quad \sum_{j=1}^{n} w_j x_j \le c, \tag{2.49}$$

$$\sum_{j \in N_k} x_j = 1, \; k = 1, \dots, r, \tag{2.50}$$

$$x_j = 0 \text{ or } 1, \; j \in N = \{1, \dots, n\} = \bigcup_{k=1}^{r} N_k, \tag{2.51}$$

assuming

$$N_h \bigcap N_k = \emptyset \quad \text{for all } h \ne k.$$

The problem is NP-hard, since any instance of KP, having r elements of profit p_j and weight w_j ($j = 1, \dots, r$) and capacity c, is equivalent to the instance of MCKP obtained by setting $n = 2r$, $p_j = w_j = 0$ for $j = r + 1, \dots, 2r$ and $N_k = \{k, r + k\}$ for $k = 1, \dots, r$.

MCKP can be solved in pseudo-polynomial time through dynamic programming as follows. Given a pair of integers l ($1 \le l \le r$) and $\hat{c}$ ($0 \le \hat{c} \le c$), consider the sub-instance of MCKP consisting of subsets $N_1, \dots, N_l$ and capacity $\hat{c}$. Let $f_l(\hat{c})$ denote its optimal solution value, i.e.

$$f_l(\hat{c}) = \max \left\{ \sum_{j \in \hat{N}} p_j x_j : \sum_{j \in \hat{N}} w_j x_j \le \hat{c}, \sum_{j \in N_k} x_j = 1 \text{ for } k = 1, \dots, l, \right.$$

$$\left. x_j = 0 \text{ or } 1 \quad \text{for } j \in \hat{N} \right\},$$

where $\hat{N} = \bigcup_{k=1}^{l} N_k$, and assume that $f_l(\hat{c}) = -\infty$ if the sub-instance has no feasible solution. Let

$$\overline{w}_k = \min\{w_j : j \in N_k\} \quad \text{for } k = 1, \dots, r;$$

clearly,

$$f_1(\hat{c}) = \begin{cases} -\infty & \text{for } \hat{c} = 0, \dots, \overline{w}_1 - 1; \\ \max\{p_j : j \in N_1, w_j \le \hat{c}\} & \text{for } \hat{c} = \overline{w}_1, \dots, c; \end{cases}$$

for $l = 2, \dots, r$ we then have

$$f_l(\hat{c}) = \begin{cases} -\infty & \text{for } \hat{c} = 0, \dots, \sum_{k=1}^{l} \overline{w}_k - 1; \\ \max\{f_{l-1}(\hat{c} - w_j) + p_j : j \in N_l, w_j \le \hat{c}\} & \\ & \text{for } \hat{c} = \sum_{k=1}^{l} \overline{w}_k, \dots, c. \end{cases}$$

The optimal solution is the state corresponding to $f_r(c)$. If we have $\sum_{k=1}^{r} \overline{w}_k > c$ then the instance has no feasible solution, and we obtain $f_r(c) = -\infty$. For each value of l, the above computation requires $O(|N_l|c)$ operations, so the overall time complexity of the method is $O(nc)$.

The execution of any algorithm for MCKP can be conveniently preceded by a reduction phase, using the following

Dominance Criterion 2.1. *For any $N_k (k = 1, \dots, r)$, if there exist two items $i, j \in N_k$ such that*

$$p_i \le p_j \quad \text{and} \quad w_i \ge w_j$$

then there exists an optimal solution to MCKP in which $x_i = 0$, i.e. item i is dominated.

Proof. Obvious from (2.50). □

As is the case for KP, dynamic programming can solve only instances of limited size. Larger instances are generally solved through branch-and-bound algorithms, based on the exact solution of the continuous relaxation of the problem, $C(MCKP)$, defined by (2.48)–(2.50) and

$$0 \le x_j \le 1, \quad j \in N. \tag{2.52}$$

An instance of $C(MCKP)$ can be further reduced through the following

Dominance Criterion 2.2. *For any $N_k (k = 1, \dots, r)$, if there exist three items $h, i, j \in N_k$ such that*

$$w_h < w_i < w_j \quad \text{and} \quad \frac{p_i - p_h}{w_i - w_h} \leq \frac{p_j - p_i}{w_j - w_i} \tag{2.53}$$

then there exists an optimal solution to $C(MCKP)$ in which $x_i = 0$, i.e. item i is dominated.

We do not give a formal proof of this criterion. However, it can be intuitively verified by representing the items of N_k as in Figure 2.8 and observing that

(i) after application of Dominance Criterion 2.1, the remaining items can only correspond to points in the shaded triangles;

(ii) for $C(MCKP)$, all points i of each triangle are dominated by the pair of vertices h, j (since for any value $x_i \neq 0$, there can be found a combination of values x_h, x_j producing a higher profit).

Hence

(iii) after application of Dominance Criterion 2.2, only those items remain which

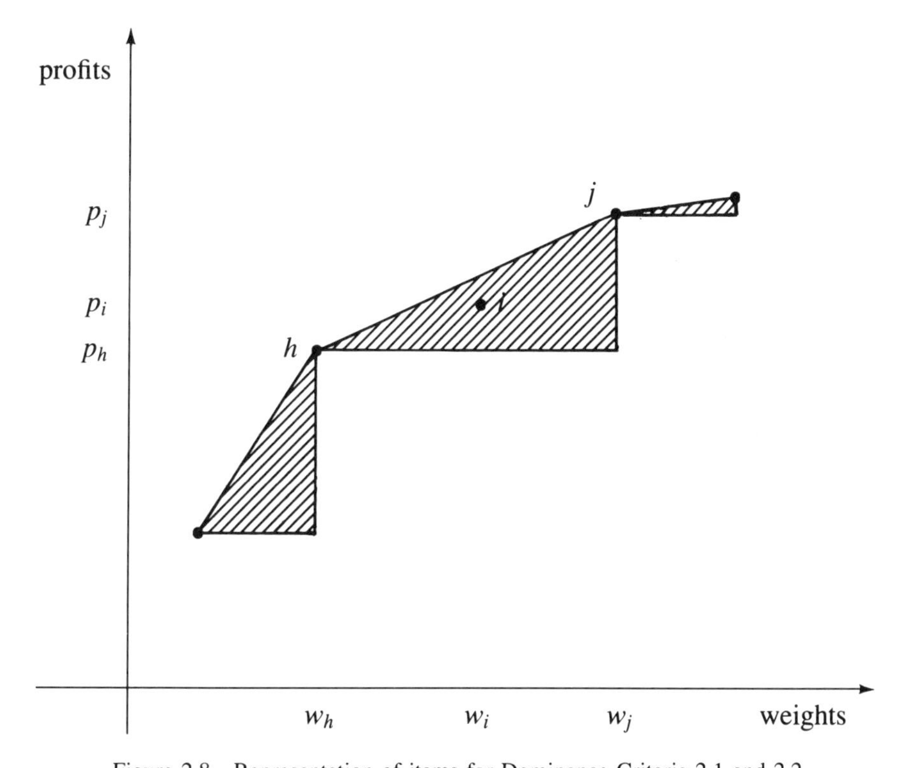

Figure 2.8 Representation of items for Dominance Criteria 2.1 and 2.2

correspond to the vertices defining the segments of the piecewise (concave) linear function.

In addition, by analysing the structure of the Linear Program corresponding to $C(MCKP)$, it is not difficult to see that

(iv) in the optimal solution of $C(MCKP)$, $r-1$ variables (corresponding to items in $r-1$ different subsets) have value 1; for the remaining subset, either one variable has value 1 or two variables (corresponding to consecutive vertices in Figure 2.8) have a fractional value.

Formal proofs of all the above properties can be found, e.g., in Sinha and Zoltners (1979).

As previously mentioned, the reduction and optimal solution of $C(MCKP)$ play a central role in all branch-and-bound algorithms for MCKP.

The reduction, based on Dominance Criteria 2.1 and 2.2, is obtained (see, e.g., Sinha and Zoltners, (1979)) by sorting the items in each subset according to increasing weights and then applying the criteria. The time complexity for this phase is clearly $O(\sum_{k=1}^{r} |N_k| \log |N_k|)$, i.e. $O(n \log \max\{|N_k| : 1 \leq k \leq r\})$.

$O(n \log r)$ algorithms for the solution of the reduced $C(MCKP)$ instance have been presented by Sinha and Zoltners (1979) and Glover and Klingman (1979). Zemel (1980) has improved the time complexity for this second phase to $O(n)$. A further improvement has been obtained by Dudzinski and Walukiewicz (1984b), who have presented an $O(r \log^2(n/r))$ algorithm.

The reduction phase is clearly the heaviest part of the process. However, in a branch-and-bound algorithm for MCKP, it is performed only at the root node, while the second phase must be iterated during execution.

Algorithms for solving $C(MCKP)$ in $O(n)$ time, without sorting and reducing the items, have been independently developed by Dyer (1984) and Zemel (1984). These results, however, have not been used, so far, in branch-and-bound algorithms for MCKP, since the reduction phase is essential for the effective solution of the problem.

Branch-and-bound algorithms for MCKP have been presented by Nauss (1978), Sinha and Zoltners (1979), Armstrong, Kung, Sinha and Zoltners (1983), Dyer, Kayal and Walker (1984), Dudzinski and Walukiewicz (1984b, 1987).

The Fortran implementation of the Dyer, Kayal and Walker (1984) algorithm can be obtained from Professor Martin E. Dyer.

3

Bounded knapsack problem

3.1 INTRODUCTION

The *Bounded Knapsack Problem* (BKP) is: given n *item types* and a *knapsack*, with

p_j = *profit* of an item of type j;

w_j = *weight* of an item of type j;

b_j = *upper bound* on the availability of items of type j;

c = *capacity* of the knapsack,

select a number x_j $(j = 1, \dots, n)$ of items of each type so as to

$$\text{maximize} \quad z = \sum_{j=1}^{n} p_j x_j \tag{3.1}$$

$$\text{subject to} \quad \sum_{j=1}^{n} w_j x_j \leq c, \tag{3.2}$$

$$0 \leq x_j \leq b_j \text{ and integer}, \quad j \in N = \{1, \dots, n\}. \tag{3.3}$$

BKP is a generalization of the 0-1 knapsack problem (Chapter 2), in which $b_j = 1$ for all $j \in N$.

We will assume, without loss of generality, that

$$p_j, w_j, b_j \text{ and } c \text{ are positive integers}, \tag{3.4}$$

$$\sum_{j=1}^{n} b_j w_j > c, \tag{3.5}$$

$$b_j w_j \leq c \text{ for } j \in N. \tag{3.6}$$

Violation of assumption (3.4) can be handled through a straightforward adaptation of the Glover (1965) method used for the 0-1 knapsack problem

(Section 2.1). If assumption (3.5) is violated then we have the trivial solution $x_j = b_j$ for all $j \in N$, while for each j violating (3.6) we can replace b_j with $\lfloor c/w_j \rfloor$. Also, the way followed in Section 2.1 to transform minimization into maximization forms can be immediately extended to BKP.

Unless otherwise specified, we will suppose that the item types are ordered so that

$$\frac{p_1}{w_1} \geq \frac{p_2}{w_2} \geq \cdots \geq \frac{p_n}{w_n} \tag{3.7}$$

A close connection between the bounded and the 0-1 knapsack problems is self-evident, so all the mathematical and algorithmic techniques analysed in Chapter 2 could be extended to the present case. The literature on BKP, however, is not comparable to that on the binary case, especially considering the last decade. The main reason for such a phenomenon is, in our opinion, the possibility of transforming BKP into an equivalent 0-1 form with a generally limited increase in the number of variables, and hence effectively solving BKP through algorithms for the 0-1 knapsack problem.

In the following sections we give the transformation technique (Section 3.2) and consider in detail some of the basic results concerning BKP (Section 3.3). The algorithmic aspects of the problem are briefly examined in Section 3.4. We do not give detailed descriptions of the algorithms since the computational results of Section 3.5 show that the last generation of algorithms for the 0-1 knapsack problem, when applied to transformed instances of BKP, outperforms the (older) specialized algorithms for the problem.

The final section is devoted to the special case of BKP in which $b_j = +\infty$ for all $j \in N$ *(Unbounded Knapsack Problem)*. For this case, interesting theoretical results have been obtained. In addition, contrary to what happens for BKP, specialized algorithms usually give the best results.

3.2 TRANSFORMATION INTO A 0-1 KNAPSACK PROBLEM

The following algorithm transforms a BKP, as defined by (3.1)–(3.3), into an equivalent 0-1 knapsack problem with

$$\begin{aligned} \hat{n} &= \text{ number of variables;} \\ (\hat{p}_j) &= \text{ profit vector;} \\ (\hat{w}_j) &= \text{ weight vector;} \\ \hat{c} = c &= \text{ capacity.} \end{aligned}$$

For each item-type j of BKP, we introduce a series of $\lfloor \log_2 b_j \rfloor$ items, whose profits and weights are, respectively, (p_j, w_j), $(2p_j, 2w_j)$, $(4p_j, 4w_j)$, ..., and one item such that the total weight (resp. profit) of the new items equals $b_j w_j$ (resp. $b_j p_j$).

```
procedure TB01:
input: n, (p_j), (w_j), (b_j);
output: n̂, (p̂_j), (ŵ_j);
begin
    n̂ := 0;
    for j := 1 to n do
        begin
            β := 0;
            k := 1;
            repeat
                if β + k > b_j then k := b_j − β;
                n̂ := n̂ + 1;
                p̂_n̂ := kp_j;
                ŵ_n̂ := kw_j;
                β := β + k;
                k := 2k
            until β = b_j
        end
end.
```

The transformed problem has $\hat{n} = \sum_{j=1}^{n} \lceil \log_2(b_j + 1) \rceil$ binary variables, hence $O(\hat{n})$ gives the time complexity of the procedure. To see that the transformed problem is equivalent to the original one, let $\hat{x}_{j_1}, \ldots, \hat{x}_{j_q}$ ($q = \lceil \log_2(b_j + 1) \rceil$) be the binary variables introduced for x_j and notice that item j_h corresponds to n_h items of type j, where

$$n_h = \begin{cases} 2^{h-1} & \text{if } h < q; \\ b_j - \sum_{i=1}^{q-1} 2^{i-1} & \text{if } h = q. \end{cases}$$

Hence $x_j = \sum_{h=1}^{q} n_h \hat{x}_{j_h}$ can take any integer value between 0 and b_j.

Notice that the transformation introduces 2^q binary combinations, i.e. $2^q - (b_j + 1)$ redundant representations of possible x_j values (the values from n_q to $2^{q-1} - 1$ have a double representation). Since, however, q is the minimum number of binary variables needed to represent the integers from 0 to b_j, any alternative transformation must introduce the same number of redundancies.

Example 3.1

Consider the instance of BKP defined by

$$\begin{aligned} n &= 3; \\ (p_j) &= (10,\ 15,\ 11); \\ (w_j) &= (\ 1,\ \ 3,\ \ 5); \\ (b_j) &= (\ 6,\ \ 4,\ \ 2); \\ c &= 10. \end{aligned}$$

Applying TB01, we get the equivalent 0-1 form:

$\hat{n} = 8;$

$(\hat{p}_j) = (10, 20, 30, 15, 30, 15, 11, 11);$

$(\hat{w}_j) = (1, 2, 3, 3, 6, 3, 5, 5).$

Items 1 to 3 correspond to the first item type, with double representation of the value $x_1 = 3$. Items 4 to 6 correspond to the second item type, with double representation of the values $x_2 = 1$, $x_2 = 2$ and $x_2 = 3$. Items 7 and 8 correspond to the third item type, with double representation of the value $x_3 = 1$. □

3.3 UPPER BOUNDS AND APPROXIMATE ALGORITHMS

3.3.1 Upper bounds

The optimal solution $\overline{x}$ of the continuous relaxation of BKP, defined by (3.1), (3.2) and

$$0 \le x_j \le b_j, \qquad j \in N,$$

can be derived in a straightforward way from Theorem 2.1. Assume that the items are sorted according to (3.7) and let

$$s = \min \left\{ j : \sum_{i=1}^{j} b_i w_i > c \right\} \tag{3.8}$$

be the *critical item type*. Then

$$\overline{x}_j = b_j \qquad \text{for } j = 1, \dots, s-1,$$

$$\overline{x}_j = 0 \qquad \text{for } j = s+1, \dots, n,$$

$$\overline{x}_s = \frac{\overline{c}}{w_s},$$

where

$$\overline{c} = c - \sum_{j=1}^{s-1} b_j w_j .$$

Hence the optimal continuous solution value is

$$\sum_{j=1}^{s-1} b_j p_j + \overline{c}\frac{p_s}{w_s},$$

and an upper bound for BKP is

$$U_1 = \sum_{j=1}^{s-1} b_j p_j + \left\lfloor \bar{c} \frac{p_s}{w_s} \right\rfloor . \tag{3.9}$$

A tighter bound has been derived by Martello and Toth (1977d) from Theorem 2.2. Let

$$z' = \sum_{j=1}^{s-1} b_j p_j + \left\lfloor \frac{\bar{c}}{w_s} \right\rfloor p_s \tag{3.10}$$

be the total profit obtained by selecting b_j items of type j for $j = 1, \dots, s-1$, and $\lfloor \bar{x}_s \rfloor$ items of type s. The corresponding residual capacity is

$$c' = \bar{c} - \left\lfloor \frac{\bar{c}}{w_s} \right\rfloor w_s .$$

Then

$$U^0 = z' + \left\lfloor c' \frac{p_{s+1}}{w_{s+1}} \right\rfloor \tag{3.11}$$

is an upper bound on the solution value we can obtain if no further items of type s are selected, while selecting at least one additional item of this type produces upper bound

$$U^1 = z' + \left\lfloor p_s - (w_s - c') \frac{p_{s-1}}{w_{s-1}} \right\rfloor . \tag{3.12}$$

Hence

$$U_2 = \max\,(U^0, U^1) \tag{3.13}$$

is an upper bound for BKP. Since from (3.9) we can write $U_1 = z' + \lfloor c' p_s / w_s \rfloor$, $U^0 \leq U_1$ is immediate, while $U^1 \leq U_1$ is proved by the same algebraic manipulations as those used in Theorem 2.2 (ii). $U_2 \leq U_1$ then follows.

The time complexity for the computation of U_1 or U_2 is $O(n)$ if the item types are already sorted. If this is not the case, the computation can still be done in $O(n)$ time through an immediate adaptation of procedure CRITICAL_ ITEM of Section 2.2.2.

Determining the continuous solution of BKP in 0-1 form still produces bound U_1. The same does not hold for U_2, since (3.11) and (3.12) explicitly consider the nature of BKP hence U^0 and U^1 are tighter than the corresponding values obtainable from the 0-1 form.

Example 3.1 (continued)

The critical item type is $s = 2$. Hence

$$U_1 = 60 + \left\lfloor 4 \frac{15}{3} \right\rfloor = 80.$$

$$U^0 = 75 + \left\lfloor 1 \frac{11}{5} \right\rfloor = 77;$$

$$U^1 = 75 + \left\lfloor 15 - 2 \frac{10}{1} \right\rfloor = 70;$$

$$U_2 = 77.$$

Considering the problem in 0-1 form and applying (2.10) and (2.16), we would obtain $U_1 = U_2 = 80$. □

Since $U_2 \le U_1 \le z' + p_s \le 2z$, the worst-case performance ratio of U_1 and U_2 is at most 2. To see that $\rho(U_1) = \rho(U_2) = 2$, consider the series of problems with $n = 3$, $p_j = w_j = k$ and $b_j = 1$ for all j, and $c = 2k - 1$: we have $U_1 = U_2 = 2k - 1$ and $z = k$, so U_1/z and U_2/z can be arbitrarily close to 2 for k sufficiently large.

All the bounds introduced in Section 2.3 for the 0-1 knapsack problem can be generalized to obtain upper bounds for BKP. This could be done either in a straightforward way, by applying the formulae of Section 2.3 to BKP in 0-1 form (as was done for U_1) or, better, by exploiting the peculiar nature of the problem (as was done for U_2). This second approach, not yet dealt with in the literature, could be a promising direction of research.

3.3.2 Approximate algorithms

Value z' defined by (3.10) is an immediate feasible solution value for BKP. Let z be the optimal solution value. Then the absolute error $z - z'$ is bounded by p_s (since $z' \le z \le U_1 \le z' + p_s$), while the ratio z'/z can be arbitrarily close to 0 (consider, e.g., $n = 2$, $p_1 = w_1 = 1$, $p_2 = w_2 = k$, $b_1 = b_2 = 1$ and $c = k$, for k sufficiently large). The worst-case performance ratio, however, can be improved to 1/2 by computing (still in $O(n)$ time)

$$z^h = \max (z', p_s)$$

as the approximate solution value. In fact, $z \le z' + p_s \le 2z^h$, and a tightness example is: $n = 2$, $p_1 = w_1 = 1$, $p_2 = w_2 = k$, $b_1 = 1$, $b_2 = 2$ and $c = 2k$, for k sufficiently large.

If the item types are sorted according to (3.7), a more effective *greedy algorithm* is the following:

```
procedure GREEDYB:
input: n, c, (p_j), (w_j), (b_j);
output: z^g, (x_j);
begin
    c̄ := c;
    z^g := 0;
```

```
    j* := 1;
    for j := 1 to n do
        begin
            x_j := min(⌊c̄/w_j⌋, b_j);
            c̄ := c̄ − w_j x_j;
            z^g := z^g + p_j x_j;
            if b_j p_j > b_j* p_j* then j* := j
        end;
    if b_j* p_j* > z^g then
        begin
            z^g := b_j* p_j*;
            for j := 1 to n do x_j := 0;
            x_j* := b_j*
        end
end.
```

The worst-case performance ratio is $\frac{1}{2}$, since trivially $z^g \geq z^h$ and the series of problems with $n = 3$, $p_1 = w_1 = 1$, $p_2 = w_2 = p_3 = w_3 = k$, $b_1 = b_2 = b_3 = 1$ and $c = 2k$ proves the tightness. The time complexity is clearly $O(n)$, plus $O(n\log n)$ for sorting.

Transforming BKP into an equivalent 0-1 problem and then applying any of the *polynomial-time*, or *fully polynomial-time approximation schemes* of Section 2.8, we obtain approximate solutions obeying the worst-case bounds defined for such schemes. In fact the two formulations of any instance have, of course, the same optimal value, and the solution determined by the scheme for the 0-1 formulation preserves feasibility and value for the bounded formulation. Hence the worst-case performance ratio is maintained. The time and space complexities of the resulting schemes are given by those in Section 2.8, with n replaced by $\hat{n} = \sum_{j=1}^{n} \lceil \log_2(b_j + 1) \rceil$.

In this case too, better results could be obtained by defining approximation schemes explicitly based on the specific structure of BKP.

3.4 EXACT ALGORITHMS

In this section we briefly outline the most important algorithms from the literature for the exact solution of BKP. The reason for not giving a detailed description of these methods is the fact that they are generally useless for effective solution of the problem. In fact, the high level of sophistication of the algorithms for the 0-1 knapsack problem has not been followed in the algorithmic approach to BKP, so the most effective way to solve bounded knapsack problems nowadays is to transform them into 0-1 form and then apply one of the algorithms of Section 2.9. (This is confirmed by the experimental results we present in the next section.) Of course, a possible direction of research could be the definition of more effective specific algorithms for BKP through adaptation of the results of Chapter 2.

3.4.1 Dynamic programming

Let $f_m(\hat{c})$ denote the optimal solution value of the sub-instance of BKP defined by item types $1, \dots, m$ and capacity $\hat{c}$ $(1 \le m \le n,\ 0 \le \hat{c} \le c)$. Clearly

$$f_1(\hat{c}) = \begin{cases} 0 & \text{for } \hat{c} = 0, \dots, w_1 - 1; \\ p_1 & \text{for } \hat{c} = w_1, \dots, 2w_1 - 1; \\ \dots & \\ (b_1 - 1)p_1 & \text{for } \hat{c} = (b_1 - 1)w_1, \dots, b_1 w_1 - 1; \\ b_1 p_1 & \text{for } \hat{c} = b_1 w_1, \dots, c. \end{cases}$$

$f_m(\hat{c})$ can then be computed, by considering increasing values of m from 2 to n, and, for each m, increasing values of $\hat{c}$ from 0 to c, as

$$f_m(\hat{c}) = \max\{ f_{m-1}(\hat{c} - l w_m) + l p_m : l \text{ integer},\ 0 \le l \le \min(b_m, \lfloor \hat{c}/w_m \rfloor)\}.$$

The optimal solution value of BKP is given by $f_n(c)$. For each m, $O(cb_m)$ operations are necessary to compute $f_m(\hat{c})$ $(\hat{c} = 0, \dots, c)$. Hence the overall time complexity for solving BKP is $O(c \sum_{m=1}^{n} b_m)$, i.e. $O(nc^2)$ in the worst case. The space complexity is $O(nc)$, since the solution vector corresponding to each $f_m(\hat{c})$ must also be stored.

The basic recursion above has been improved on, among others, by Gilmore and Gomory (1966) and Nemhauser and Ullmann (1969). Dynamic programming, however, can only solve problems of very limited size. (Nemhauser and Ullmann (1969) report that their algorithm required 74 seconds to solve, on an IBM-7094, a problem instance with $n = 50$ and $b_j = 2$ for each j.)

3.4.2 Branch-and-bound

Martello and Toth (1977d) adapted procedure MT1 of Section 2.5.2 to BKP. The resulting depth-first branch-and-bound algorithm, which incorporates upper bound U_2 of Section 3.3.1, is not described here, but could easily be derived from procedure MTU1 presented in Section 3.6.2 for the unbounded knapsack problem. (See also a note by Aittoniemi and Oehlandt (1985).)

Ingargiola and Korsh (1977) presented a reduction algorithm related to the one in Ingargiola and Korsh (1973) (Section 2.7) and imbedded it into a branch-search algorithm related to the one in Greenberg and Hegerich (1970) (Section 2.5). (See also a note by Martello and Toth (1980c).)

Bulfin, Parker and Shetty (1979) have proposed a different branch-and-bound strategy, incorporating penalties in order to improve the bounding phase.

Aittoniemi (1982) gives an experimental comparison of the above algorithms, indicating the Martello and Toth (1977d) one as the most effective. As already mentioned, however, all these methods are generally outperformed by algorithm MT2 (Section 2.9.3) applied to the transformed 0-1 instance. The Fortran implementation of this algorithm (MTB2) is included in the present volume.

3.5 COMPUTATIONAL EXPERIMENTS

In Tables 3.1, 3.2 and 3.3 we analyse the experimental behaviour of exact and approximate algorithms for BKP through data sets similar to those used for the 0-1 knapsack problem, i.e.:

uncorrelated: p_j and w_j uniformly random in [1,1000];

weakly correlated: w_j uniformly random in [1,1000], p_j uniformly random in $[w_j - 100, w_j + 100]$;

strongly correlated: w_j uniformly random in [1,1000], $p_j = w_j + 100$.

For all data sets, the values b_j are uniformly random in [5,10], and c is set to $0.5 \sum_{j=1}^{n} b_j w_j$ (so about half of the items are in the optimal solution).

The tables compare the Fortran IV implementations of the following methods:

Table 3.1 Uncorrelated problems: p_j and w_j uniformly random in [1,1000], b_j uniformly random in [5,10]; $c = 0.5 \sum_{j=1}^{n} b_j w_j$. HP 9000/840 in seconds. Average times (average percentage errors) over 20 problems

	MTB	IK	MTB2	MTB2 approximate	GREEDYB
n	time	time	time	time (% error)	time (% error)
25	0.034	0.022	0.023	0.011(0.09851)	0.001(0.09721)
50	0.121	0.115	0.049	0.020(0.04506)	0.005(0.04775)
100	0.464	0.149	0.084	0.031(0.02271)	0.012(0.01354)
200	1.761	0.462	0.143	0.061(0.01166)	0.023(0.00809)
500	9.705	5.220	0.395	0.158(0.00446)	0.065(0.00246)
1 000	36.270	11.288	0.583	0.324(0.00079)	0.138(0.00071)
2 000	88.201	33.490	1.107	0.649(0.00097)	0.272(0.00033)
5 000	159.213	106.550	2.272	1.585(0.00028)	0.745(0.00008)
10 000	—	—	3.599	3.055(0.00031)	1.568(0.00003)
20 000	—	—	6.689	6.195(0.00011)	3.332(0.00001)
30 000	—	—	9.445	9.692(0.00010)	5.144(0.00000)
40 000	—	—	14.119	13.443(0.00003)	7.080(0.00000)
50 000	—	—	14.836	15.298(0.00005)	8.942(0.00000)

Table 3.2 Weakly correlated problems: w_j uniformly random in [1,1000], p_j in $[w_j - 100, w_j + 100]$, b_j uniformly random in [5,10]; $c = 0.5 \sum_{j=1}^{n} b_j w_j$. HP 9000/840 in seconds. Average times (average percentage errors) over 20 problems

n	MTB time	IK time	MTB2 time	MTB2 approximate time (% error)	GREEDYB time (% error)
25	0.051	0.206	0.075	0.012(0.08072)	0.001(0.13047)
50	0.150	0.855	0.199	0.019(0.03975)	0.007(0.04214)
100	0.478	3.425	0.207	0.037(0.01384)	0.014(0.01374)
200	1.350	8.795	0.354	0.061(0.00901)	0.021(0.00461)
500	6.232	25.840	0.532	0.147(0.00414)	0.057(0.00126)
1 000	16.697	59.182	0.574	0.292(0.00228)	0.125(0.00054)
2 000	39.707	57.566	0.810	0.568(0.00242)	0.265(0.00015)
5 000	131.670	131.212	1.829	1.572(0.00062)	0.725(0.00004)
10 000	—	—	3.359	3.052(0.00037)	1.572(0.00001)
20 000	—	—	6.973	6.633(0.00021)	3.293(0.00000)
30 000	—	—	9.785	9.326(0.00016)	5.089(0.00000)
40 000	—	—	6435.178	12.182(0.00017)	6.966(0.00000)
50 000	—	—	—	15.473(0.00010)	8.533(0.00000)

Table 3.3 Strongly correlated problems: w_j uniformly random in [1,1000], $p_j = w_j + 100$, b_j uniformly random in [5,10]; $c = 0.5 \sum_{j=1}^{n} b_j w_j$. HP 9000/840 in seconds. Average times (average percentage errors) over 20 problems

n	MTB time	IK time	MTB2 time	MTB2 approximate time (% error)	GREEDYB time (% error)
25	3.319	216.864	23.091	0.012(0.36225)	0.002(0.62104)
50	279.782	—	4513.810	0.018(0.14509)	0.005(0.22967)
100	—	—	—	0.037(0.14295)	0.010(0.16482)
200	—	—	—	0.066(0.07570)	0.023(0.08262)
500	—	—	—	0.139(0.03866)	0.059(0.03919)
1 000	—	—	—	0.283(0.01688)	0.123(0.01701)
2 000	—	—	—	0.589(0.00818)	0.265(0.00822)
5 000	—	—	—	1.529(0.00352)	0.756(0.00352)
10 000	—	—	—	3.133(0.00181)	1.558(0.00181)
20 000	—	—	—	5.794(0.00064)	3.169(0.00064)
30 000	—	—	—	9.847(0.00054)	5.065(0.00054)
40 000	—	—	—	12.058(0.00042)	6.705(0.00042)
50 000	—	—	—	15.265(0.00034)	8.603(0.00034)

exact algorithms:

MTB = Martello and Toth (1977d);

IK = Ingargiola and Korsh (1977);

MTB2 = Transformation through procedure TB01 (Section 3.2) and solution through algorithm MT2 (Section 2.9.3);

approximate algorithms:

MTB2 approximate = MTB2 with heuristic version of MT2 (Section 2.10.2);

GREEDYB = greedy algorithm (Section 3.3.2).

All runs were executed on an HP 9000/840 (with option "-o" for the Fortran compiler), with values of n ranging from 25 to 50 000 (for $n > 50\,000$, the size of the transformed instances could exceed the memory limit). The tables give average times and percentage errors computed over sets of 20 instances each. The errors are computed as $100(z - z^a)/z$, where z^a is the approximate solution value, and z either the optimal solution value (when available) or upper bound U_2 introduced in Section 3.3.1. The execution of each algorithm was halted as soon as the average time exceeded 100 seconds.

MTB2 is clearly the most efficient exact algorithm for uncorrelated and weakly correlated problems. Optimal solution of strongly correlated problems appears to be practically impossible. As for the heuristic algorithms, GREEDYB dominates the approximate version of MTB2 for uncorrelated and weakly correlated problems, but produces higher errors for strongly correlated problems with $n \leq 2\,000$. The anomalous entry in Table 3.2 (MTB2 exact, $n = 40\,000$) was produced by an instance requiring more than 34 hours!

3.6 A SPECIAL CASE: THE UNBOUNDED KNAPSACK PROBLEM

In this section we consider the problem arising from BKP when an unlimited number of items of each type is available, i.e. the *Unbounded Knapsack Problem* (UKP)

$$\text{maximize} \quad z = \sum_{j=1}^{n} p_j x_j \tag{3.14}$$

$$\text{subject to} \quad \sum_{j=1}^{n} w_j x_j \leq c, \tag{3.15}$$

$$x_j \geq 0 \text{ and integer}, \quad j \in N = \{1, \ldots, n\}. \tag{3.16}$$

The problem remains NP-hard, as proved in Lueker (1975) by transformation from subset-sum. However, it can be solved in polynomial time in the $n = 2$ case (Hirschberg and Wong (1976), Kannan (1980)). Notice that the result is not trivial, since a naive algorithm, testing $x_1 = i$, $x_2 = \lfloor (c - iw_1)/w_2 \rfloor$ for i taking on integer values from 0 to $\lfloor c/w_1 \rfloor$, would require a time $O(c)$, exponential in the input length.

UKP can clearly be formulated (and solved) by defining an equivalent BKP with $b_j = \lfloor c/w_j \rfloor$ for $j = 1, \dots, n$, but algorithms for BKP generally perform rather poorly in instances of this kind. Also transformation into an equivalent 0-1 knapsack problem is possible (through a straightforward adaptation of the method of Section 3.2), but usually impractical since the number of resulting binary variables $(\sum_{j=1}^{n} \lceil \log_2(\lfloor c/w_j \rfloor + 1) \rceil)$ is generally too elevated for practical solution of the problem.

We maintain assumptions (3.4) and (3.7), while (3.6) transforms into

$$w_j \leq c \qquad \text{for } j \in N \tag{3.17}$$

and (3.5) is satisfied by any instance of UKP.

3.6.1 Upper bounds and approximate algorithms

The optimal solution of the continuous relaxation of UKP, defined by (3.14), (3.15) and

$$x_j \geq 0, \qquad j \in N,$$

is $\overline{x}_1 = c/w_1$, $\overline{x}_j = 0$ for $j = 2, \dots, n$, and provides the trivial upper bound

$$U_0 = \left\lfloor c \frac{p_1}{w_1} \right\rfloor .$$

By also imposing $\overline{x}_1 \leq \lfloor c/w_1 \rfloor$, which must hold in any integer solution, the continuous solution is

$$\overline{x}_1 = \left\lfloor \frac{c}{w_1} \right\rfloor ,$$

$$\overline{x}_j = 0 \qquad \text{for } j = 3, \dots, n,$$

$$\overline{x}_2 = \frac{\overline{c}}{w_2},$$

where

$$\overline{c} = c (\text{mod } w_1). \tag{3.18}$$

This provides the counterpart of upper bound U_1 of Section 3.3.1, i.e.

$$U_1 = \left\lfloor \frac{c}{w_1} \right\rfloor p_1 + \left\lfloor \bar{c}\frac{p_2}{w_2} \right\rfloor. \tag{3.19}$$

(Note that the critical item type is always $s = 2$.)

The counterpart of the improved upper bound U_2 is

$$U_2 = \max\,(U^0, U^1), \tag{3.20}$$

where

$$z' = \left\lfloor \frac{c}{w_1} \right\rfloor p_1 + \left\lfloor \frac{\bar{c}}{w_2} \right\rfloor p_2, \tag{3.21}$$

$$c' = \bar{c} (\mathrm{mod}\ w_2), \tag{3.22}$$

$$U^0 = z' + \left\lfloor c'\frac{p_3}{w_3} \right\rfloor, \tag{3.23}$$

$$U^1 = z' + \left\lfloor p_2 - (w_2 - c')\frac{p_1}{w_1} \right\rfloor. \tag{3.24}$$

In this case, however, we can exploit the fact that $s = 2$ to obtain a better bound. Remember (see Section 3.3.1) that U^1 is an upper bound on the solution value we can obtain if at least $\lfloor \bar{c}/w_2 \rfloor + 1$ items of type 2 are selected. Notice now that this can be done only if at least $\lceil (w_2 - c')/w_1 \rceil$ items of type 1 are removed from the solution corresponding to z', and that $c' + \lceil (w_2 - c')/w_1 \rceil w_1$ units of capacity are then available for the items of type 2. Hence, a valid upper bound can be obtained by replacing U^1 with

$$\overline{U}^1 = z' + \left\lfloor \left(c' + \left\lceil \frac{w_2 - c'}{w_1} \right\rceil w_1\right) \frac{p_2}{w_2} - \left\lceil \frac{w_2 - c'}{w_1} \right\rceil p_1 \right\rfloor. \tag{3.25}$$

Furthermore, $\overline{U}^1 \leq U^1$ since, with $c' + \lceil (w_2 - c')/w_1 \rceil w_1 \geq w_2$, $\overline{U}^1$ is obtained by "moving" a greater number of capacity units from items of type 1 to (worse) items of type 2. We have thus proved the following

Theorem 3.1 (Martello and Toth, 1990a)

$$U_3 = \max\,(U^0, \overline{U}^1), \tag{3.26}$$

where U^0 and $\overline{U}^1$ are defined by (3.18), (3.21)–(3.23) *and* (3.25), *is an upper bound for UKP and, for any instance, $U_3 \leq U_2$.*

The time complexity for the computation of U_0, U_1, U_2 and U_3 is $O(n)$, since only the three largest ratios p_j/w_j are needed.

Example 3.2

Consider the instance of UKP defined by

$$n = 3;$$
$$(p_j) = (20, 5, 1);$$
$$(w_j) = (10, 5, 3);$$
$$c = 39.$$

The upper bounds are

$$U_0 = 78.$$
$$U_1 = 60 + \left\lfloor 9\frac{5}{5} \right\rfloor = 69.$$
$$U^0 = 65 + \left\lfloor 4\frac{1}{3} \right\rfloor = 66;$$
$$U^1 = 65 + \left\lfloor 5 - 1\frac{20}{10} \right\rfloor = 68;$$
$$U_2 = 68.$$
$$\overline{U}^1 = 65 + \left\lfloor \left(4 + \left\lceil \frac{1}{10} \right\rceil 10\right)\frac{5}{5} - \left\lceil \frac{1}{10} \right\rceil 20 \right\rfloor = 59;$$
$$U_3 = 66. \square$$

Since $U_3 \leq U_2 \leq U_1 \leq U_0 \leq z' + p_1 \leq 2z$, the worst-case performance ratio of all bounds is at most 2. To see that $\rho(U_0) = \rho(U_1) = \rho(U_2) = \rho(U_3) = 2$, consider the series of problems with $n = 3$, $p_j = w_j = k$ for all j, and $c = 2k - 1$: we have $U_0 = U_1 = U_2 = U_3 = 2k - 1$ and $z = k$, so the ratio (upper bound)/z can be arbitrarily close to 2 for k sufficiently large.

The heuristic solution value defined by (3.21) has an interesting property. Remember that the analogous values z' defined for BKP (Section 3.3.2) and for the 0-1 knapsack problem (Section 2.4) can provide an arbitrarily bad approximation of the optimal value z. For any instance of UKP, instead, we have it that $z'/z \geq \frac{1}{2}$. The proof is immediate by observing that $z - z' \leq p_1$ and, from (3.17), $z' \geq p_1$. The series of problems with $n = 2$, $p_1 = w_1 = k + 1$, $p_2 = w_2 = k$ and $c = 2k$ shows that $\frac{1}{2}$ is tight, since $z'/z = (k + 1)/(2k)$ can be arbitrarily close to $\frac{1}{2}$ for k sufficiently large. Also notice that the same property holds for the simpler heuristic value $z'' = \lfloor c/w_1 \rfloor p_1$.

The greedy algorithm of Section 3.3.2 can now be simplified as follows. (We assume that the item types are sorted according to (3.7).)

```
procedure GREEDYU:
input: n, c, (p_j), (w_j);
output: z^g, (x_j);
begin
    c̄ := c;
    z^g := 0;
    for j := 1 to n do
        begin
            x_j := ⌊c̄/w_j⌋;
            c̄ := c̄ − w_j x_j;
            z^g := z^g + p_j x_j
        end
end.
```

The time complexity of GREEDYU is $O(n)$, plus $O(n\log n)$ for the preliminary sorting.

Magazine, Nemhauser and Trotter (1975) studied theoretical properties of the greedy algorithm when applied to the minimization version of UKP. In particular, they determined necessary and sufficient conditions for the optimality of the greedy solution (see also Hu and Lenard (1976) for a simplified proof), and analysed the worst-case absolute error produced by the algorithm. Ibarra and Kim (1975) adapted their fully polynomial-time approximation scheme for the 0-1 knapsack problem (Section 2.8.2) to UKP. The resulting scheme produces, for any fixed $\varepsilon > 0$, a solution having worst-case relative error not greater than ε in time $O(n + (1/\varepsilon^4)\log(1/\varepsilon))$ and space $O(n + (1/\varepsilon^3))$. Also Lawler (1979) derived from his algorithm for the 0-1 knapsack problem (Section 2.8.2) a fully polynomial-time approximation scheme for UKP, obtaining time and space complexity $O(n+(1/\varepsilon^3))$.

3.6.2 Exact algorithms

An immediate recursion for computing the dynamic programming function $f_m(\hat{c})$ (see Section 3.4.1), is

$$f_1(\hat{c}) = \left\lfloor \frac{\hat{c}}{w_1} \right\rfloor p_1 \qquad \text{for } \hat{c} = 0, \dots, c;$$

$$f_m(\hat{c}) = \max\left\{ f_{m-1}(\hat{c} - lw_m) + lp_m : l \text{ integer}, \ 0 \le l \le \left\lfloor \frac{\hat{c}}{w_m} \right\rfloor \right\}$$
$$\text{for } m = 2, \dots, n \text{ and } \hat{c} = 0, c.$$

The time complexity for determining $z = f_n(c)$ is $O(nc^2)$.

Gilmore and Gomory (1965) have observed that a better recursion for computing $f_m(\hat{c})$, for $m = 2, \dots, n$, is

$$f_m(\hat{c}) = \begin{cases} f_{m-1}(\hat{c}) & \text{for } \hat{c} = 0, \dots, w_m - 1; \\ \max\ (f_{m-1}(\hat{c}), f_m(\hat{c} - w_m) + p_m) & \text{for } \hat{c} = w_m, \dots, c, \end{cases}$$

which reduces the overall time complexity to $O(nc)$.

Specialized dynamic programming algorithms for UKP have been given by Gilmore and Gomory (1966), Hu (1969), Garfinkel and Nemhauser (1972), Greenberg and Feldman (1980), Greenberg (1985, 1986). Dynamic programming, however, is usually capable of solving only instances of limited size.

More effective algorithms, based on branch-and-bound, have been proposed by Gilmore and Gomory (1963), Cabot (1970) and Martello and Toth (1977d). The last one has proved to be experimentally the most effective (Martello and Toth, 1977d), and derives from algorithm MT1 for the 0-1 knapsack problem, described in Section 2.5.2. Considerations (i) to (iii) of that section easily extend to this algorithm, while parametric computation of upper bounds (consideration (iv)) is no longer needed, since the current critical item type is always the next item type to be considered. The general structure of the algorithm and the variable names used in the following detailed description are close to those in MT1. It is assumed that the item types are sorted according to (3.7).

procedure MTU1:
input: $n, c, (p_j), (w_j)$;
output: $z, (x_j)$;
begin
1. [initialize]
 $z := 0$;
 $\hat{z} := 0$;
 $\hat{c} := c$;
 $p_{n+1} := 0$;
 $w_{n+1} := +\infty$;
 for $k := 1$ **to** n **do** $\hat{x}_k := 0$;
 compute the upper bound $U = U_3$ on the optimal solution value;
 for $k := n$ **to** 1 **step** -1 **do** compute $m_k = \min\{w_i : i > k\}$;
 $j := 1$;
2. [build a new current solution]
 while $w_j > \hat{c}$ **do**
  if $z \geq \hat{z} + \lfloor \hat{c} p_{j+1} / w_{j+1} \rfloor$ **then go to** 5 **else** $j := j + 1$;
 $y := \lfloor \hat{c} / w_j \rfloor$;
 $u := \lfloor (\hat{c} - y w_j) p_{j+1} / w_{j+1} \rfloor$;
 if $z \geq \hat{z} + y p_j + u$ **then go to** 5;
 if $u = 0$ **then go to** 4;
3. [save the current solution]
 $\hat{c} := \hat{c} - y w_j$;
 $\hat{z} := \hat{z} + y p_j$;
 $\hat{x}_j := y$;
 $j := j + 1$;

```
        if ĉ ≥ m_{j-1} then go to 2;
        if z ≥ ẑ then go to 5;
        y := 0;
4. [update the best solution so far]
        z := ẑ + y p_j;
        for k := 1 to j - 1 do x_k := x̂_k;
        x_j := y;
        for k := j + 1 to n do x_k := 0;
        if z = U then return ;
5. [backtrack]
        find i = max{k < j : x̂_k > 0};
        if no such i then return ;
        ĉ := ĉ + w_i;
        ẑ := ẑ - p_i;
        x̂_i := x̂_i - 1;
        if z ≥ ẑ + ⌊ĉ p_{i+1}/w_{i+1}⌋ then
            begin
                comment: remove all items of type i;
                ĉ := ĉ + w_i x̂_i;
                ẑ := ẑ - p_i x̂_i;
                x̂_i := 0;
                j := i;
                go to 5
            end;
        j := i + 1;
        if ĉ - w_i ≥ m_i then go to 2;
        h := i;
6. [try to replace one item of type i with items of type h]
        h := h + 1;
        if z ≥ ẑ + ⌊ĉ p_h/w_h⌋ then go to 5;
        if w_h = w_i then go to 6;
        if w_h > w_i then
            begin
                if w_h > ĉ or z ≥ ẑ + p_h then go to 6;
                z := ẑ + p_h;
                for k := 1 to n do x_k := x̂_k;
                x_h := 1;
                if z = U then return;
                i := h;
                go to 6
            end
        else
            begin
                if ĉ - w_h < m_{h-1} then go to 6;
                j := h;
                go to 2
            end
end.
```

Example 3.3

Consider the instance of UKP defined by

$n = 7$;

$(p_j) = (20, 39, 52, 58, 31, 4, 5)$;

$(w_j) = (15, 30, 41, 46, 25, 4, 5)$;

$c = 101$.

Figure 3.1 gives the decision-tree produced by algorithm MTU1. □

The Fortran implementation of procedure MTU1 is included in that of procedure MTU2, which is described in the next section.

3.6.3 An exact algorithm for large-size problems

Experimental results with algorithm MTU1, reported in Martello and Toth (1977b), show a behaviour close to that of analogous algorithms for the 0-1 knapsack problem, i.e.: (i) in spite of its worst-case complexity, many instances of UKP can be exactly solved within reasonable computing times, even for very large values of n; (ii) when this is possible, the sorting time is usually a very large fraction of the total time; however, (iii) only the item types with the highest values of the ratio p_j/w_j are selected for the solution, i.e. $\max\{j : x_j > 0\} \ll n$.

The concept of core problem (Section 2.9) can be extended to UKP by recalling that, in this case, the critical item type is always the second one. Hence, given a UKP and supposing, without loss of generality, that $p_j/w_j > p_{j+1}/w_{j+1}$ for $j = 1, \dots, n-1$, we define the *core* as

$$C = \{1, 2, \dots, \overline{n} \equiv \max\{j : x_j > 0\}\},$$

and the *core problem* as

$$\text{maximize } z = \sum_{j \in C} p_j x_j$$

$$\text{subject to } \sum_{j \in C} w_j x_j \leq c,$$

$$x_j \geq 0 \text{ and integer}, \qquad j \in C.$$

If we knew "a priori" the value of $\overline{n}$, we could solve UKP by setting $x_j = 0$ for all j such that $p_j/w_j < p_{\overline{n}}/w_{\overline{n}}$, determining C as $\{j : p_j/w_j \geq p_{\overline{n}}/w_{\overline{n}}\}$ and solving the resulting core problem by sorting only the items in C. $\overline{n}$ cannot, of course, be "a priori" identified, but we can determine an approximate core without sorting as follows.

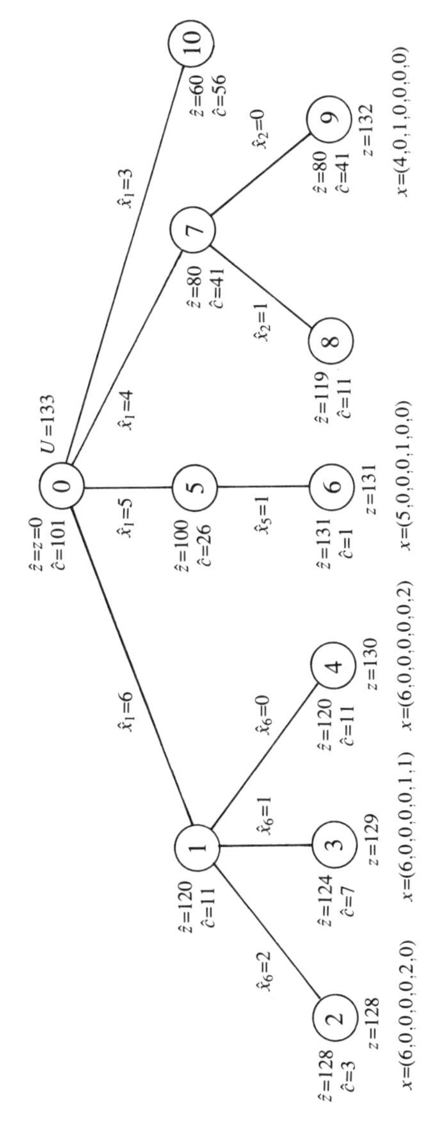

Figure 3.1 Decision-tree of procedure MTU1 for Example 3.3

Assuming no condition on the ratios p_j/w_j, we select a tentative value for $p_{\overline{n}}/w_{\overline{n}}$ and solve the corresponding core problem: if the solution value equals that of an upper bound, then we have the optimum; otherwise, we reduce the variables not in the core and, if any variables are left, we try again with a decreased tentative value. Reduction is based on the following criterion. Let $U_q(j)$ denote upper bound U_q ($q = 1, 2$ or 3) of Section 3.6.1 for UKP, with the additional constraint $x_j = 1$, i.e. an upper bound on the solution value that UKP can have if item type j is used for the solution. If, for j not in the approximate core, we have $U_q(j) \leq z$ (where z denotes the solution value of the approximate core problem), then we know that x_j must take the value 0 in any solution better than the current one. Given a tentative value ϑ for the initial core problem size, the resulting algorithm is thus (Martello and Toth, 1990a) the following.

procedure MTU2:
input: $n, c, (p_j), (w_j), \vartheta$;
output: $z, (x_j)$;
begin
  $k := 0$;
  $\overline{N} := \{1, 2, \dots, n\}$;
  repeat
    $k := \min(k + \vartheta, |\overline{N}|)$;
    find the kth largest value r in $\{p_j/w_j : j \in \overline{N}\}$;
    $G := \{j \in \overline{N} : p_j/w_j > r\}$;
    $E := \{j \in \overline{N} : p_j/w_j = r\}$;
    $\overline{E}$:= any subset of E such that $|\overline{E}| = k - |G|$;
    $C := G \cup \overline{E}$;
    sort the item types in C according to decreasing p_j/w_j ratios;
    exactly solve the core problem, using MTU1, and let z and (x_j) define
      the solution;
    if $k = \vartheta$ (**comment**: first iteration) **then**
      compute upper bound U_3 of Section 3.6.1;
    if $z < U_3$ **then** (**comment**: reduction)
      for each $j \in \overline{N} \backslash C$ **do**
        begin
          $u := U_1(j)$;
          if $u > z$ **then** $u := U_3(j)$;
          if $u \leq z$ **then** $\overline{N} := \overline{N} \backslash \{j\}$
        end
  until $z = U_3$ or $\overline{N} = C$;
  for each $j \in \{1, \dots, n\} \backslash C$ **do** $x_j := 0$
end.

At each iteration, the exact solution of the core problem is obtained by first identifying *dominated* item types in C, then applying algorithm MTU1 to the undominated item types. Dominances are identified as follows.

Definition 3.1 *Given an instance of UKP, relative to item types set N, item type*

$k \in N$ is dominated if the optimal solution value does not change when k is removed from N.

Theorem 3.2 (Martello and Toth, 1990a) *Given any instance of UKP and an item type k, if there exists an item type j such that*

$$\left\lfloor \frac{w_k}{w_j} \right\rfloor p_j \geq p_k \tag{3.27}$$

then k is dominated.

Proof. Given a feasible solution in which $x_k = \alpha > 0$ and $x_j = \beta$, a better solution can be obtained by setting $x_k = 0$ and $x_j = \beta + \lfloor w_k/w_j \rfloor \alpha$. In fact: (i) the new solution is feasible, since $\lfloor w_k/w_j \rfloor \alpha w_j \leq \alpha w_k$; (ii) the profit produced by item type j in the new solution is no less than that produced by item types j and k in the given solution, since, from (3.27), $\lfloor w_k/w_j \rfloor \alpha p_j \geq \alpha p_k$. □

Corollary 3.1 *All dominated item types can be efficiently eliminated from the core as follows:*

1. *sort the item types according to (3.7), breaking ties so that $w_j \leq w_{j+1}$;*
2. **for** $j := 1$ **to** $|C| - 1$ **do**
   for $k := j + 1$ **to** $|C|$ **do if** (3.27) *holds* **then** $C := C \backslash \{k\}$.

Proof. Condition (3.27) never holds if either $p_j/w_j < p_k/w_k$ or $w_k < w_j$. □

Hence the time complexity to eliminate the dominated item types is $O(|C|^2)$ (or $O(n^2)$, if the original UKP is considered).

Example 3.3 (continued)

Taking $\vartheta = 4$, the core problem is defined by:

$(p_j) = (20, 39, 52, 58)$;

$(w_j) = (15, 30, 41, 46)$.

Applying Corollary 3.1, we find that item type 1 dominates item types 2 and 4. Applying MTU1 to the resulting problem, defined by

$(p_j) = (20, 52)$;

$(w_j) = (15, 41)$,

we obtain the branch-decision tree of Figure 3.2.

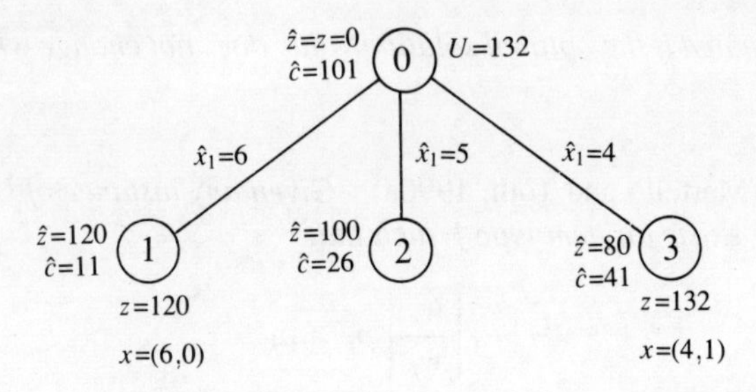

Figure 3.2 Decision-tree of procedure MTU2 for Example 3.3

The core problem solution value ($z = 132$) is not equal to upper bound U_3 relative to the original instance without the dominated item types ($U_3 = \max(120 + \lfloor 11 \frac{31}{25} \rfloor,\ 120 + \lfloor (11 + \lceil \frac{30}{15} \rceil 15) \frac{52}{41} - \lceil \frac{30}{15} \rceil 20 \rfloor) = 133$). Hence we apply the reduction phase:

$$j = 5 : U_1(5) = 31 + \left(100 + \left\lfloor 1 \frac{52}{41} \right\rfloor \right) = 132 \le z;$$

$$j = 6 : U_1(6) = 4 + \left(120 + \left\lfloor 7 \frac{52}{41} \right\rfloor \right) = 132 \le z;$$

$$j = 7 : U_1(7) = 5 + \left(120 + \left\lfloor 6 \frac{52}{41} \right\rfloor \right) = 132 \le z.$$

Since all the item types not in core are reduced, we conclude that the core problem has produced the optimal solution $z = 132$, $(x_j) = (4, 0, 1, 0, 0, 0, 0)$. □

The initial tentative value ϑ was experimentally determined as

$$\vartheta = \max \left(100, \left\lfloor \frac{n}{100} \right\rfloor \right).$$

The Fortran implementation of algorithm MTU2 is included in the present volume.

3.6.4 Computational experiments

Table 3.4 compares the algorithms for UKP on the same data sets of Section 3.5, but with w_j uniformly randomly generated in the range [10,1000], so as to avoid the occurrence of trivial instances in which the item type with largest p_j/w_j ratio has $w_j = 1$ (so $x_1 = c$ is the optimal solution).

For all problems, c was set to $0.5 \sum_{j=1}^{n} w_j$ for $n \le 100\,000$, to $0.1 \sum_{j=1}^{n} w_j$ (in order to avoid integer overflows) for $n > 100\,000$.

We compare the Fortran IV implementations of algorithms MTU1 and MTU2. The kth largest ratio p_j/w_j was determined through the algorithm given in Fischetti and Martello (1988) (including Fortran implementation). All runs have been

Table 3.4 w_j uniformly random in [10,1000]; $c = 0.5\sum_{j=1}^{n} w_j$ for $n \leq 100\,000$, $c = 0.1\sum_{j=1}^{n} w_j$ for $n > 100\,000$. HP 9000/840 in seconds. Average times over 20 problems

		Uncorrelated: p_j unif. random in [1,1000]		Weakly correlated: p_j unif. random in $[w_j - 100, w_j + 100]$		Strongly correlated: $p_j = w_j + 100$	
n	Sorting	MTU1	MTU2	MTU1	MTU2	MTU1	MTU2
50	0.01	0.01	0.01	0.01	0.01	0.01	0.01
100	0.01	0.01	0.01	0.01	0.01	0.01	0.01
200	0.02	0.02	0.02	0.02	0.03	0.06	0.02
500	0.05	0.05	0.04	0.05	0.05	131.70	0.04
1 000	0.11	0.11	0.07	0.11	0.08	—	0.08
2 000	0.24	0.24	0.13	0.24	0.14	—	0.14
5 000	0.60	0.68	0.32	0.62	0.29	—	0.35
10 000	1.35	1.44	0.60	1.37	0.66	—	0.62
20 000	3.21	3.31	1.23	494.93	1.18	—	1.39
30 000	4.71	5.38	1.82	—	1.94	—	1.91
40 000	6.12	9.67	2.73	—	2.48	—	2.66
50 000	8.04	21.91	3.25	—	3.30	—	3.34
60 000	10.53	41.11	3.90	—	3.71	—	4.10
70 000	12.50	17.63	4.89	—	4.50	—	4.81
80 000	13.86	172.00	5.28	—	5.00	—	5.12
90 000	15.56	—	5.88	—	5.41	—	5.68
100 000	17.96	—	5.83	—	6.22	—	5.81
150 000	27.41	—	10.05	—	10.14	—	9.95
200 000	37.56	—	13.08	—	11.98	—	13.26
250 000	48.55	—	17.35	—	17.52	—	17.94

executed on an HP 9000/840 with option "-o" for the Fortran compiler. For each data set and value of n, Table 3.4 gives the average running times (including sorting), expressed in seconds, computed over 20 problem instances. Sorting times are also separately shown. Execution of an algorithm was halted as soon as the average running time exceeded 100 seconds.

The table shows that MTU2 always dominates MTU1, and can solve very large problems with reasonable computing time also in the case of strongly correlated data sets. The initial value of ϑ always produced the optimal solution. With the exception of strongly correlated data sets, MTU1 requires negligible extra computational effort after sorting, when $n \leq 10\,000$. For larger values of n, the branch-and-bound phase can become impractical. This shows that the superiority of MTU2 (particularly evident for very large instances and for strongly correlated problems) derives not only from the avoided sorting phase but also from application of the dominance criterion. In fact, the number of undominated item types was always very small and almost independent of n.

4

Subset-sum problem

4.1 INTRODUCTION

The *Subset-Sum Problem* (SSP) is: given a set of n *items* and a *knapsack*, with

$$w_j = \textit{weight} \text{ of item } j;$$

$$c = \textit{capacity} \text{ of the knapsack,}$$

select a subset of the items whose total weight is closest to, without exceeding, c, i.e.

$$\text{maximize} \quad z = \sum_{j=1}^{n} w_j x_j \tag{4.1}$$

$$\text{subject to} \quad \sum_{j=1}^{n} w_j x_j \leq c, \tag{4.2}$$

$$x_j = 0 \text{ or } 1, \qquad j \in N = \{1, \ldots, n\}, \tag{4.3}$$

where

$$x_j = \begin{cases} 1 & \text{if item } j \text{ is selected;} \\ 0 & \text{otherwise.} \end{cases}$$

The problem is related to the *diophantine equation*

$$\sum_{j=1}^{n} w_j x_j = \hat{c}, \tag{4.4}$$

$$x_j = 0 \text{ or } 1, \qquad j = 1, \ldots, n, \tag{4.5}$$

in the sense that the optimal solution value of SSP is the largest $\hat{c} \leq c$ for which (4.4)–(4.5) has a solution.

SSP, which is also called the *Value Independent Knapsack Problem* or *Stickstacking Problem,* is a particular case of the 0-1 knapsack problem (Chapter 2)—arising when $p_j = w_j$ for all j—hence, without loss of generality, we will assume that

$$w_j \text{ and } c \text{ are positive integers,} \tag{4.6}$$

$$\sum_{j=1}^{n} w_j > c, \tag{4.7}$$

$$w_j < c \text{ for } j \in N. \tag{4.8}$$

Violation of such assumptions can be handled as indicated in Section 2.1.

The problem arises in situations where a quantitative target should be reached, such that its negative deviation (or loss of, e.g., trim, space, time, money) must be minimized and a positive deviation is not allowed. Recently, massive SSP's have been used in several coefficient reduction procedures for strengthening LP bounds in general integer programming (see Dietrich and Escudero (1989a, 1989b)).

SSP can obviously be solved (either exactly or heuristically) by any of the methods described in Chapter 2 for the 0-1 knapsack problem. It deserves, however, specific treatment since specialized algorithms usually give much better results. A macroscopic reason for this is the fact that all upper bounds of Sections 2.2 and 2.3 give, for SSP, the trivial value c (since $p_j/w_j = 1$ for all j). SSP can be seen, in fact, as the extreme case of correlation between profits and weights (see Section 2.10). As a consequence, one would even expect catastrophic behaviour of the branch-and-bound algorithms for the 0-1 knapsack problem, degenerating, for SSP, into complete enumeration (because of the value c produced, at all decision nodes, by upper bound computations). This is not always true. In fact, as soon as a feasible solution of value c is determined, one can obviously stop execution and, as we will see, this phenomenon often occurs for problems in which the number of items is not too small. Also note that the reduction procedures of Section 2.7 have no effect on SSP, because of the bound's uselessness.

We describe exact and approximate algorithms for SSP in Sections 4.2 and 4.3, respectively, and analyse computational results in Section 4.4.

4.2 EXACT ALGORITHMS

4.2.1 Dynamic programming

Given a pair of integers m $(1 \leq m \leq n)$ and $\hat{c}$ $(0 \leq \hat{c} \leq c)$, let $f_m(\hat{c})$ be the optimal solution value of the sub-instance of SSP consisting of items $1, \dots, m$ and capacity $\hat{c}$. The dynamic programming recursion for computing $f_n(c)$ (optimal solution value of SSP) can be easily derived from that given in Section 2.6 for the 0-1 knapsack problem:

$$f_1(\hat{c}) = \begin{cases} 0 & \text{for } \hat{c} = 0, \dots, w_1 - 1; \\ w_1 & \text{for } \hat{c} = w_1, \dots, c; \end{cases}$$

for $m = 2, \ldots, n$:

$$f_m(\hat{c}) = \begin{cases} f_{m-1}(\hat{c}) & \text{for } \hat{c} = 0, \ldots, w_m - 1; \\ \max\left(f_{m-1}(\hat{c}), f_{m-1}(\hat{c} - w_m) + w_m\right) & \text{for } \hat{c} = w_m, \ldots, c. \end{cases}$$

The time and space complexity to compute $f_n(c)$ is thus $O(nc)$.

Faaland (1973) has presented a specialized dynamic programming approach of the same complexity, which is also suitable for the *bounded* version of SSP, defined by (4.1), (4.2) and

$$0 \leq x_j \leq b_j, \qquad j = 1, \ldots, n,$$

$$x_j \text{ integer}, \qquad j = 1, \ldots, n.$$

The algorithm derives from a recursive technique given by Verebriusova (1904) to determine the number of non-negative integer solutions to diophantine equations (4.4).

Ahrens and Finke (1975) proposed a more effective approach which reduces, on average, the time and space required to solve the problem. The method derives from their dynamic programming algorithm for the 0-1 knapsack problem (Section 2.6.2) and makes use of the "replacement selection" technique, described in Knuth (1973), in order to combine the partial lists obtained by partitioning the variables into four subsets.

Because of the large core memory requirements (the Ahrens and Finke (1975) algorithm needs about $2^{n/4+4}$ words) dynamic programming can be used only for small instances of the problem.

Martello and Toth (1984a) used "partial" dynamic programming lists to obtain a hybrid algorithm (described in the next section) to effectively solve also large instances of SSP. These lists are obtained through a recursion conceptually close to procedure REC2 given in Section 2.6.1 for the 0-1 knapsack problem, but considering only states of total weight not greater than a given value $\tilde{c} \leq c$. The particular structure of SSP produces considerable simplifications. The undominated states are in this case those corresponding to values of $\hat{c}$ for which the diophantine equation (4.4)–(4.5) has a solution. At stage m, the undominated states are determined from the following information, relative to the previous stage:

$$s = \text{number of states at the previous stage}; \tag{4.9}$$

$$b = 2^{m-1}; \tag{4.10}$$

$$W1_i = \text{ total weight of the } i\text{th state } (i = 1, \ldots, s); \tag{4.11}$$

$$X1_i = \{x_1, x_2, \ldots, x_{m-1}\} \qquad \text{for } i = 1, \ldots, s, \tag{4.12}$$

where x_j defines the value of the jth variable in the solution relative to the ith state, i.e. $W1_i = \sum_{j=1}^{m-1} w_j x_j$. Vector $W1_i$ is assumed to be ordered according to strictly increasing values. The procedure updates values (4.9) and (4.10), and stores the new values of (4.11) and (4.12) in $(W2_k)$ and $(X2_k)$. Sets $X1_i$ and $X2_k$ are encoded as bit strings. Note that, for SSP, states having the same weight are equivalent, i.e. dominating each other. In such situations, the algorithm stores only one state, so vector $(W2_k)$ results are ordered according to strictly increasing values. On input, it is assumed that $W1_0 = X1_0 = 0$.

```
procedure RECS:
input: s, b, (W1_i), (X1_i), w_m, c̃;
output: s, b, (W2_k), (X2_k);
begin
    i := 0;
    k := 0;
    h := 1;
    y := w_m;
    W1_{s+1} := +∞;
    W2_0 := 0;
    X2_0 := 0;
    while min(y, W1_h) ≤ c̃ do
        begin
            k := k + 1;
            if W1_h ≤ y then
                begin
                    W2_k := W1_h;
                    X2_k := X1_h;
                    h := h + 1
                end
            else
                begin
                    W2_k := y;
                    X2_k := X1_i + b
                end
            if W2_k = y then
                begin
                    i := i + 1;
                    y := W1_i + w_m
                end
        end
    s := k;
    b := 2b
end.
```

Procedure RECS is a part of the hybrid algorithm described in the next section. It can also be used, however, to directly solve SSP as follows.

```
procedure DPS:
input: n, c, (w_j);
output: z, (x_j);
begin
    c̃ := c;
    W1_0 := 0;
    X1_0 := 0;
    s := 1;
    b := 2;
    W1_1 := w_1;
    X1_1 := 1;
    m := 2;
    repeat
        call RECS;
        rename W2 and X2 as W1 and X1, respectively;
        m := m + 1
    until m > n or W1_s = c;
    z := W1_s;
    determine (x_j) by decoding X1_s
end.
```

The time complexity of RECS is $O(s)$. Since s is bounded by min $(2^m - 1, \tilde{c})$, the time complexity of DPS is $O(\min(2^{n+1}, nc))$.

4.2.2 A hybrid algorithm

Martello and Toth (1984a) used a combination of dynamic programming and tree-search to effectively solve SSP. Assume that the items are sorted beforehand so that

$$w_1 \geq w_2 \geq \cdots \geq w_n. \tag{4.13}$$

The algorithm starts by applying the dynamic programming recursion to a subset containing the last (small) items and by storing the corresponding state lists. Tree-search is then performed on the remaining (large) items. In this way, the state weights in the lists are small and close to each other, while, in the branch-decision tree, the current residual capacity $\hat{c}$ takes small values after few forward moves, allowing use of the dynamic programming lists.

The algorithm starts by determining two partial state lists:

(i) given a prefixed value $MA < n - 1$, list (WA_i, XA_i), $i = 1, \dots, SA$, contains all the undominated states induced by the last MA items;

(ii) given two prefixed values MB $(MA < MB < n)$ and $\overline{c}$ $(w_n < \overline{c} < c)$, list (WB_i, XB_i), $i = 1, \dots, SB$, contains the undominated states of weight not greater than $\overline{c}$ induced by the last MB items.

Figure 4.1, in which $NA = n - MA + 1$ and $NB = n - MB + 1$, shows the states covered by the two lists: the thick lines approximate the step functions giving, for each item, the maximum state weight obtained at the corresponding iteration.

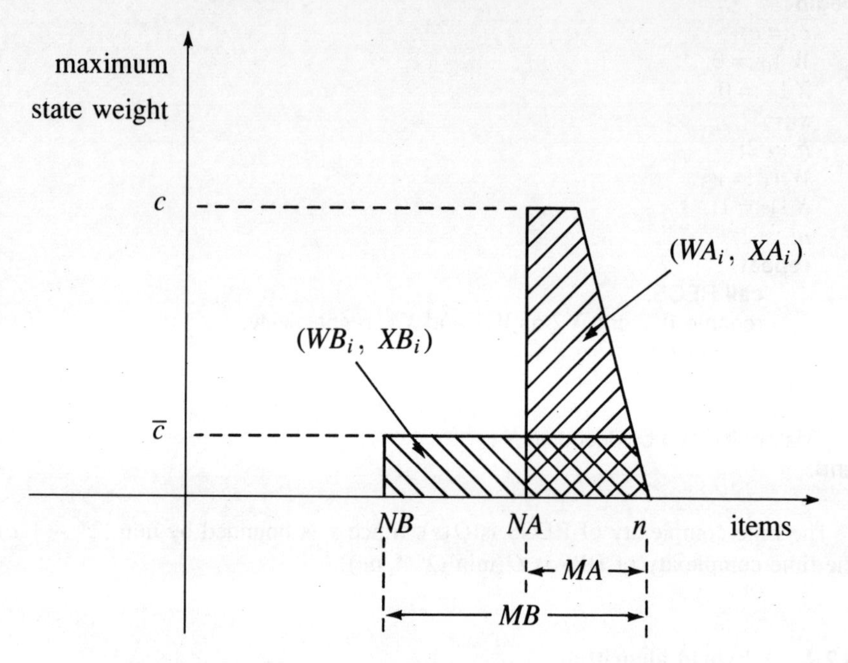

Figure 4.1 States covered by the dynamic programming lists

The following procedure determines the two lists. List (WA_i, XA_i) is first determined by calling procedure RECS in reverse order, i.e. determining, for $m = n, n-1, \ldots, NA (\equiv n - MA + 1)$, the optimal solution value $\varphi_m(\hat{c})$ of the sub-instance defined by items $m, m+1, \ldots, n$ and capacity $\hat{c} \leq c$. List (WB_i, XB_i) is then initialized to contain those states of (WA_i, XA_i) whose weight is not greater than $\overline{c}$, and completed by calling RECS for $m = NA - 1, NA - 2, \ldots, NB (\equiv n - MB + 1)$. Note that the meaning of XA and XB is consequently altered with respect to (4.12).

procedure LISTS:
input: $n, c, (w_j), NA, NB, \overline{c}$;
output: $SA, (WA_i), (XA_i), SB, (WB_i), (XB_i)$;
begin

comment: determine list (WA_i, XA_i);
$\tilde{c} := c$;
$W1_0 := 0$;
$X1_0 := 0$;
$s := 1$;
$b := 2$;
$W1_1 := w_n$;
$X1_1 := 1$;

```
    m := n - 1;
    repeat
        call RECS;
        rename W2 and X2 as W1 and X1, respectively;
        m := m - 1
    until m < NA or W1_s = c;
    for i := 1 to s do
        begin
            WA_i := W1_i;
            XA_i := X1_i
        end;
    SA := s;
    if WA_SA < c then (comment: determine list (WB_i, XB_i))
        begin
            c~ := c̄;
            determine, through binary search, ī = max{i : WA_i ≤ c̄};
            s := ī;
            repeat
                call RECS;
                rename W2 and X2 as W1 and X1, respectively;
                m := m - 1
            until m < NB;
            rename W1 and X1 as WB and XB, respectively;
            SB := s
        end
end.
```

Example 4.1

Consider the instance of SSP defined by

$n = 10;$

$(w_j) = (41, 34, 21, 20, 8, 7, 7, 4, 3, 3);$

$c = 50;$

MA $= 4;$

MB $= 6;$

$\overline{c} = 12.$

Calling LISTS, we obtain $SA = 9$, $SB = 8$ and the values given in Figure 4.2. □

We can now state the overall algorithm. After having determined the dynamic programming lists, the algorithm generates a binary decision-tree by setting $\hat{x}_j$ to 1 or 0 for $j = 1, \ldots, NA - 1$. Only the first $NA - 1$ items are considered, since all the feasible combinations of items $NA, \ldots, n$ are in list (WA_i, XA_i). A forward

i	WA_i	XA_i(decoded)	WB_i	XB_i(decoded)
0	0	0	0	0
1	3	1	3	1
2	4	100	4	100
3	6	11	6	11
4	7	101	7	101
5	10	111	8	100000
6	11	1100	10	111
7	13	1011	11	1100
8	14	1101	12	100100
9	17	1111		

Figure 4.2 Dynamic programming lists for Example 4.1

move starting from an item j consists in: (a) finding the first item $j' \geq j$ which can be added to the current solution; (b) adding to the current solution a feasible sequence $j', j'+1, \dots, j''$ of consecutive items until the residual capacity $\hat{c}$ is no greater than $\overline{c}$. A backtracking step consists in removing from the current solution that item j''' which was inserted last and in performing a forward move starting from $j''' + 1$.

At the end of a forward move, we determine the maximum weight δ of a dynamic programming state which can be added to the current solution. This is done by assuming the existence of two functions, A and B, to determine, respectively,

$$A(\hat{c}) = \max_{0 \leq i \leq SA} \{i : WA_i \leq \hat{c}\},$$

$$B(\hat{c}, j) = \max_{0 \leq i \leq SB} \{i : WB_i \leq \hat{c} \text{ and } y_k^i = 0 \text{ for all } k < j\},$$

where (y_k^i) denotes the binary vector encoded in XB_i. (Both $A(\hat{c})$ and $B(\hat{c}, j)$ can be implemented through binary search.) After updating of the current optimal solution z ($z := \max(z, (c - \hat{c}) + \delta)$), we proceed to the next forward move, unless we find that the solution values of all the descendent decision nodes are dominated by $(c - \hat{c}) + \delta$. This happens when either the next item which we could insert is one of the MA last items, or is one of the MB last items and the residual capacity $\hat{c}$ is no greater than $\overline{c}$.

Values $F_k = \sum_{j=k}^{n} w_j$ $(k = 1, \dots, n)$ are used to avoid forward moves when $\hat{c} \geq F_{j'}$ or an upper bound on the optimal solution obtainable from the move is no greater than the value of the best solution so far.

```
procedure MTS:
input: n, c, (w_j), c̄, MA, MB;
output: z, (x_j);
begin
1. [initialize]
    NA := n − MA + 1;
    NB := n − MB + 1;
    call LISTS;
    z := WA_SA;
    for k := 1 to NA − 1 do x_k := 0;
    let (y_k) be the binary vector encoded in XA_SA;
    for k := NA to n do x_k := y_k;
    if z = c then return;
    for k := n to 1 step −1 do compute F_k = Σ_{j=k}^{n} w_j;
    ẑ := 0;
    ĉ := c;
    for k := 1 to n do x̂_k := 0;
    j := 1;
2. [try to avoid the next forward move]
    while w_j > ĉ and j < NA do j := j + 1;
    if j = NA then go to 4;
    if F_j ≤ ĉ then
        begin
            if ẑ + F_j > z then (comment: new optimal solution)
                begin
                    z := ẑ + F_j;
                    for k := 1 to j − 1 do x_k := x̂_k;
                    for k := j to n do x_k := 1;
                    if z = c then return
                end;
            go to 5
        end;
    determine, through binary search, r = min{k > j : F_k ≤ ĉ};
    s := n − r + 1;
    comment: at most s items can be added to the current solution;
    u := F_j − F_{j+s};
    comment: u = Σ_{k=j}^{j+s−1} w_j = total weight of the s largest available items;
    if ẑ + u ≤ z then go to 5;
3. [perform a forward move]
    while w_j ≤ ĉ and j < NA and ĉ > c̄ do
        begin
            ĉ := ĉ − w_j;
            ẑ := ẑ + w_j;
            x̂_j := 1;
            j := j + 1
        end;
```

```
4. [use the dynamic programming lists]
   if ĉ ≤ c̄ then
       begin
           δ := WB_B(ĉ,j);
           flag := "b"
       end
   else
       begin
           δ := WA_A(ĉ);
           flag := "a";
           if δ < c̄ and z < ẑ + c̄ then
               begin
                   δ := WB_B(ĉ,j);
                   flag := "b"
               end
       end;
   comment: δ is the maximum additional weight obtainable from the lists;
   if ẑ + δ > z then (comment: update the optimal solution)
       begin
           z := ẑ + δ;
           for k := 1 to j − 1 do x_k := x̂_k;
           if flag = "a" then
               begin
                   for k := j to NA − 1 do x_k := 0;
                   let (y_k) be the vector encoded in XA_A(ĉ);
                   for k := NA to n do x_k := y_k
               end
           else
               begin
                   for k := j to NB − 1 do x_k := 0;
                   let (y_k) be the vector encoded in XB_B(ĉ,j);
                   for k := max(NB, j) to n do x_k := y_k
               end;
           if z = c then return
       end;
   if (ĉ < w_NA−1 or j = NA) then go to 5;
   if (ĉ < w_NB−1 or j ≥ NB) and (ĉ < c̄) then go to 5
   else go to 2;
5. [backtrack]
   find i = max{k < j : x̂_k = 1};
   if no such i then return;
   ĉ := ĉ + w_i;
   ẑ := ẑ − w_i;
   x̂_i := 0;
   j := i + 1;
   go to 2
end.
```

Example 4.1 (continued)

Executing MTS, we obtain:

NA = 7,

NB = 5,

(F_k) = (148, 107, 73, 52, 32, 24, 17, 10, 6, 3),

the dynamic programming lists of Figure 4.2 and the branch-decision tree of Figure 4.3. □

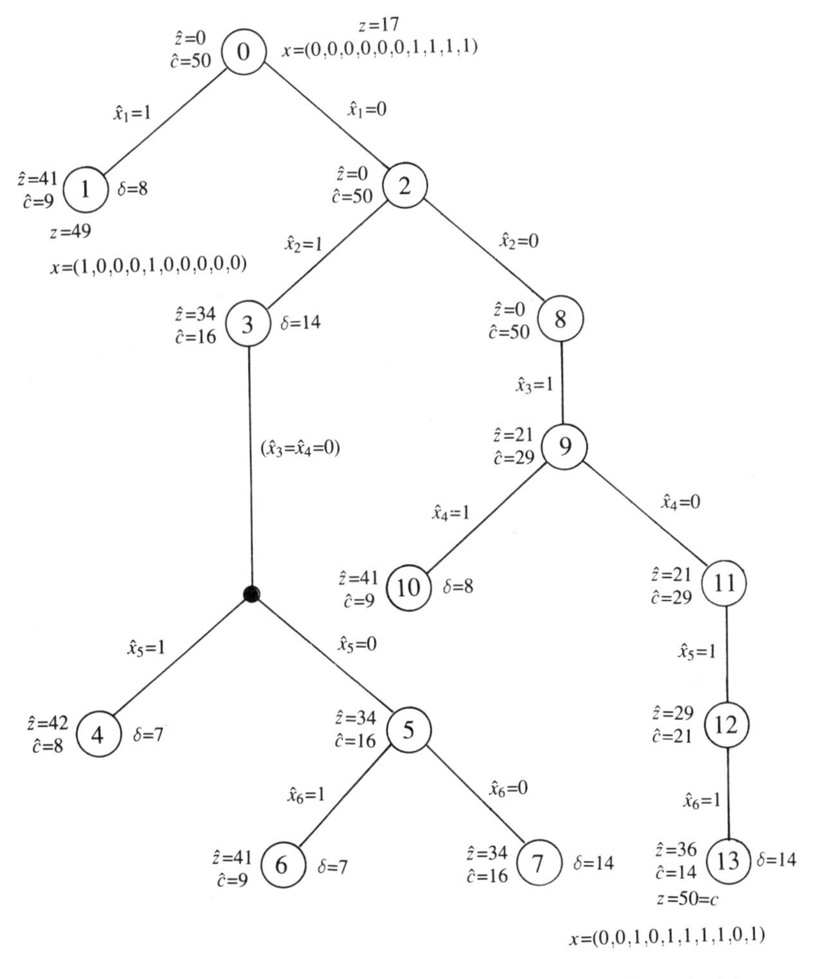

Figure 4.3 Branch-decision tree of Procedure MTS for Example 4.1

The Fortran implementation of procedure MTS is included in that of procedure MTSL, which is described in the next section. The parameters for the dynamic programming lists must take into account the "difficulty" of the problem. They have been experimentally determined as the following functions of n and $wmax = \max\{w_j\}$:

$$MA = \min(2\log_{10} wmax,\ 0.7n);$$

$$MB = \min(2.5\log_{10} wmax,\ 0.8n);$$

$$\overline{c} = 1.3 w_{NB}.$$

These values are automatically decreased by the code corresponding to MTS whenever the space required from the lists is larger than the available core memory.

A different hybrid algorithm for SSP can be found in Plateau and Elkihel (1985).

4.2.3 An algorithm for large-size problems

Computational experiments with algorithm MTS show (Martello and Toth, 1984a) that many instances of SSP can be exactly solved in reasonable computing time, since they admit a large number of feasible solutions of value c (i.e. optimal). Hence, for large-size problems, there is the possibility of finding one such solution by considering (and sorting) only a relatively small subset of the items. This can be obtained by defining a *core problem* which has a structure similar to that introduced for the 0-1 knapsack problem (Section 2.9) but can be determined much more efficiently as follows. Given an instance of SSP, we determine the *critical item* $s = \min\{j : \sum_{i=1}^{j} w_i > c\}$ and, for a prefixed value $\vartheta > 0$, we define the *core problem*

$$\text{maximize} \quad \tilde{z} = \sum_{j=s-\vartheta}^{s+\vartheta} w_j x_j \tag{4.14}$$

$$\text{subject to} \quad \sum_{j=s-\vartheta}^{s+\vartheta} w_j x_j \le \tilde{c} = c - \sum_{j=1}^{s-\vartheta-1} w_j, \tag{4.15}$$

$$x_j = 0 \text{ or } 1, \qquad j = s-\vartheta, \ldots, s+\vartheta. \tag{4.16}$$

Then we sort items $s-\vartheta, \ldots, s+\vartheta$ according to (4.13) and solve the core problem through procedure MTS. If the solution value found is equal to $\tilde{c}$ then we have an optimal solution of value c for SSP, defined by values $x_{s-\vartheta}, \ldots, x_{s+\vartheta}$ returned by MTS, and by $x_j = 1$ for $j < s-\vartheta$, $x_j = 0$ for $j > s+\vartheta$. Otherwise, we enlarge the core problem by increasing ϑ and repeat.

```
procedure MTSL:
input: n, c, (w_j), ϑ, MA, MB, c̄;
output: z, (x_j);
begin
    determine s = min{j : Σ_{i=1}^{j} w_i > c};
    repeat
        a := max(1, s − ϑ);
        b := min(n, s + ϑ);
        c̃ := c − Σ_{j=1}^{a−1} w_j;
        sort items a, a + 1, ..., b according to decreasing weights;
        call MTS for the core problem (4.14)–(4.16) and let z̃ be the solution
          value returned;
        ϑ := 2ϑ
    until z̃ = c̃ or b − a + 1 = n;
    let y_j (j = a, ..., b) be the solution vector returned by MTS;
    for j := 1 to a − 1 do x_j := 1;
    for j := a to b do x_j := y_j;
    for j := b + 1 to n do x_j := 0;
    z := z̃ + (c − c̃)
end.
```

A "good" input value for ϑ was experimentally determined as

$$\vartheta = 45.$$

The Fortran implementation of MTSL is included in the present volume.

4.3 APPROXIMATE ALGORITHMS

4.3.1 Greedy algorithms

The most immediate approach to the heuristic solution of SSP is the *Greedy Algorithm,* which consists in examining the items in any order and inserting each new item into the knapsack if it fits. By defining $p_j = w_j$ for all j, we can use procedure GREEDY given in Section 2.4 for the 0-1 knapsack problem. This procedure will consider, for SSP, the item of maximum weight alone as a possible alternative solution, and guarantee a worst-case performance ratio equal to $\frac{1}{2}$. No sorting being needed, since (2.7) is satisfied by any instance of SSP, the time complexity decreases from $O(n\log n)$ to $O(n)$.

For SSP, better average results can be obtained by sorting the items according to decreasing weights. Since in this way the item of maximum weight is always considered first (and hence inserted), we no longer need to explicitly determine it, so a considerably simpler procedure is the following. We assume that the items are sorted according to (4.13).

```
procedure GS:
input: n, c, (w_j);
output: z^g, (x_j);
begin
    ĉ := c;
    for j := 1 to n do
        if w_j > ĉ then x_j := 0
        else
            begin
                x_j := 1;
                ĉ := ĉ − w_j
            end;
    z^g := c − ĉ
end.
```

The worst-case performance ratio is still $\frac{1}{2}$, while the time complexity grows to $O(n \log n)$ because of the required sorting.

An $O(n^2)$ greedy algorithm, with better worst-case performance ratio was given by Martello and Toth (1984b). The idea is to apply the greedy algorithm n times, by considering item sets $\{1, \ldots, n\}$, $\{2, \ldots, n\}$, $\{3, \ldots, n\}$, and so on, respectively, and take the best solution. Assuming that the items are sorted according to (4.13), the algorithm is the following.

```
procedure MTGS:
input: n, c, (w_j);
output: z^g, X^h;
begin
    z^g := 0;
    for i := 1 to n do
        begin
            ĉ := c;
            Y := Ø;
            for j := i to n do
                if w_j ≤ ĉ then
                    begin
                        ĉ := ĉ − w_j;
                        Y := Y ∪ {j}
                    end;
            if c − ĉ > z^g then
                begin
                    z^g := c − ĉ;
                    X^h := Y;
                    if z^g = c then return
                end
        end
end.
```

The time complexity of MTGS is clearly $O(n^2)$. Its worst-case performance ratio is established by the following

Theorem 4.1 (Martello and Toth, 1984b) $r(\text{MTGS}) = \frac{3}{4}$.

Proof. We will denote by $z(k)$ the value $c - \hat{c}$ of the solution found by the algorithm at the kth iteration, i.e. by considering item set $\{k, \ldots, n\}$. Let

$$q = \max \{j : \exists\ k < j \text{ such that item } j \text{ is not selected for } z(k)\}, \tag{4.17}$$

$$Q = \sum_{j=q+1}^{n} w_j, \tag{4.18}$$

and note that, because of (4.17), items $q + 1, \ldots, n$ are selected for all $z(k)$ with $k \leq q + 1$. Let $z = \sum_{j=1}^{n} w_j x_j^*$ be the optimal solution value and define $A = \{j \leq q : x_j^* = 1\}$.

(a) If $|A| \leq 2$ then $z^g = z$. In fact: (i) if $|A| = 1$, with $A = \{j_1\}$, we have $z^g \geq z(j_1) \geq w_{j_1} + Q = z$; (ii) if $|A| = 2$, with $A = \{j_1, j_2\}$ and $j_1 < j_2$, we have $z^g \geq z(j_1) \geq w_{j_1} + w_{j_2} + Q = z$.

(b) If $|A| > 2$ then $z^g \geq \frac{3}{4}c \geq \frac{3}{4}z$. In fact: (i) if $w_q > \frac{1}{4}c$, we have $z^g \geq z(q-2) = w_{q-2} + w_{q-1} + w_q + Q > \frac{3}{4}c$; (ii) if $w_q \leq \frac{1}{4}c$, there must exist an iteration $\overline{k} \leq q - 1$ in which item q is not selected for $z(\overline{k})$ since $w_q > \hat{c} \geq c - z(\overline{k})$, and hence we have $z^g \geq z(\overline{k}) > c - w_q \geq \frac{3}{4}c$.

To prove that value $\frac{3}{4}$ is tight, consider the series of instances with $n = 4$, $w_1 = 2R$, $w_2 = R + 1$, $w_3 = w_4 = R$ and $c = 4R$. The optimal solution value is $z = 4R$, while $z(1) = z(2) = 3R + 1$, $z(3) = 2R$ and $z(4) = R$, so $z^g = 3R + 1$. Hence the ratio z^g/z can be arbitrarily close to $\frac{3}{4}$, for R sufficiently large. □

Note that, for the series of instances above, the optimal solution would have been produced by a modified version of the algorithm applying the greedy search at iteration k to item set $\{k, \ldots, n, 1, \ldots, k-1\}$ (the result would be $z^g = z(3) = 4R$). However, adding a fifth element with $w_5 = 1$ gives rise to a series of problems whereby z^g/z tends to $\frac{3}{4}$ for the modified algorithm as well. Also, from the practical point of view, computational experiments with the modified algorithm (Martello and Toth, 1985a) show very marginal average improvements with considerably higher computing times.

More accurate approximate solutions can obviously be obtained by using any of the approximation schemes described for the 0-1 knapsack problem (Section 2.8). However, by exploiting the special structure of the problem, we can obtain better schemes—both from the theoretical and the practical point of view—for the approximate solution of SSP.

4.3.2 Polynomial-time approximation schemes

The first polynomial-time approximation scheme was given by Johnson (1974). The idea is to identify a subset of "large" items (according to a given parameter k) and to find the corresponding optimal solution. This is completed by applying the greedy algorithm, for the residual capacity, to the remaining items. The algorithm can be efficiently implemented as the following procedure (slightly different from the original one presented by Johnson (1974)), in which k is supposed to be a positive integer:

```
procedure J(k):
input: n, c, (w_j);
output: z^h, X^h;
begin
    L := {j : w_j > c/(k + 1)};
    determine X^h ⊆ L such that z^h = Σ_{j∈X^h} w_j is closest to, without exceeding,
      c;
    ĉ := c − z^h;
    S := {1, ..., n}\L;
    sort the items in S according to decreasing weights and let m = min_{j∈S} {w_j};
    while S ≠ Ø and ĉ ≥ m do
        begin
            let j be the first item in S;
            S := S\{j};
            if w_j ≤ ĉ then
                begin
                    ĉ := ĉ − w_j;
                    X^h := X^h ∪ {j}
                end
        end;
        z^h := c − ĉ
end.
```

The time complexity to determine the initial value of z^h and the corresponding X^h through complete enumeration is $O(n^k)$, since $|X^h| \leq k$. The remaining part of the algorithm requires time $O(n\log n)$—for sorting—plus $O(n)$. The overall time complexity of J(k) is thus $O(n\log n)$ for $k = 1$, and $O(n^k)$ for $k > 1$. The space complexity is $O(n)$.

Theorem 4.2 (Johnson, 1974) $r(\text{J}(k)) = k/(k+1)$.

Proof. Let $z = \sum_{j\in X^*} w_j$ be the optimal solution value and consider partition of the optimal item set X^* into $L^* = \{j \in X^* : w_j > c/(k+1)\}$ and $S^* = X^*\backslash L^*$. Similarly, partition item set X^h returned by J(k) into $L^h = \{j \in X^h : w_j > c/(k+1)\}$ and $S^h = X^h\backslash L^h$. Since L^h is the optimal subset of $L = \{j : w_j > c/(k+1)\}$, initially determined by the algorithm, we have $\sum_{j \in L^h} w_j \geq \sum_{j \in L^*} w_j$. Hence, if $S^* \subseteq S^h$, we also have $\sum_{j \in S^h} w_j \geq \sum_{j \in S^*} w_j$ and the solution found by the algorithm is optimal. Otherwise, let $q \in S^*$ be any item not selected for S^h: it must be $w_q + z^h > c$, so $z^h > c - w_q \geq ck/(k+1) \geq zk/(k+1)$.

Tightness of the $k/(k+1)$ bound is proved by the series of problems with $n = k+2$, $w_1 = R+1$, $w_j = R$ for $j > 1$ and $c = (k+1)$. The optimal solution value is $z = (k+1)R$. Since it results that $L = \{1\}$, the heuristic solution is $z^h = kR + 1$, so the ratio z^h/z can be arbitrarily close to $k/(k+1)$ for R sufficiently large. □

Note that J(1) produces the greedy solution. In fact $L = \{j : w_j > c/2\}$, so only one item (the one with largest weight) will be selected from L while, for the remaining items, a greedy search is performed.

Example 4.2

Consider the instance of SSP defined by

$$n = 9;$$

$$(w_j) = (81, 80, 43, 40, 30, 26, 12, 11, 9);$$

$$c = 100.$$

MTGS gives, in $O(n^2)$ time: $z^h = \max(93, 92, 95, 96, 88, 58, 32, 20, 9) = 96$, $X^h = \{4, 5, 6\}$.

J(1) (as well as GS) gives, in $O(n\log n)$ time: $L = \{1, 2\}$, $z^h = 93$, $X^h = \{1, 7\}$.

J(2) gives, in $O(n^2)$ time: $L = \{1, 2, 3, 4\}$, $z^h = 95$, $X^h = \{3, 4, 7\}$.

J(3) gives, in $O(n^3)$ time: $L = \{1, 2, 3, 4, 5, 6\}$, $z^h = 99$, $X^h = \{3, 5, 6\}$.

The optimal solution $z = 100$, $X = \{2, 8, 9\}$ is found by J(11). □

A better polynomial-time approximation scheme has been found by Martello and Toth (1984b) by combining the idea in their algorithm MTGS of the previous section with that in the Sahni (1975) scheme for the 0-1 knapsack problem (see Section 2.8.1). For $k = 2$, the resulting scheme applies MTGS to the original problem (for $k = 1$ the scheme is not defined but it is assumed to be the greedy algorithm). For $k = 3$, it imposes each item in turn and applies MTGS to the resulting subproblem, taking the best solution. For $k = 4$, all possible item pairs are imposed, and so on. It will be shown in Section 4.4.2 that, for practical purposes, $k = 2$ or 3 is enough for obtaining solutions very close to the optimum. It is assumed that the items are sorted according to (4.13).

procedure MTSS(k):
input: $n, c, (w_j)$;
output: z^h, X^h;
begin
    $z^h := 0$;

```
for each M ⊂ N = {1, ..., n} such that |M| ≤ k − 2 do
    begin
        g := Σ_{j∈M} w_j;
        if g ≤ c then
            begin
                call MTGS for the subproblem defined by item set
                N\M and reduced capacity c − g, and let z^g =
                Σ_{j∈V} w_j (V ⊆ N\M) be the solution found;
                if z^g > z^h then
                    begin
                        z^h := z^g;
                        X^h := M ∪ V;
                        if z^h = c then return
                    end
            end
    end
end.
```

Since there are $O(n^{k-2})$ subsets $M \subset N$ of cardinality not greater than $k-2$, and recalling that MTGS requires $O(n^2)$ time, the overall time complexity of MTSS(k) is $O(n^k)$. The space complexity is clearly $O(n)$. From Theorem 4.1 we have $r(\text{MTSS}(2)) = \frac{3}{4}$. Martello and Toth (1984b) have proved that $r(\text{MTSS}(3)) = \frac{6}{7}$ and $(k+3)/(k+4) \le r(\text{MTSS}(k)) \le k(k+1)/(k(k+1)+2)$ for $k \ge 4$. Fischetti (1986) exactly determined the worst-case performance ratio of the scheme:

Theorem 4.3 (Fischetti, 1986) $r(\text{MTSS}(k)) = (3k-3)/(3k-2)$.

Proof. We omit the part proving that $r(\text{MTSS}(k)) \ge (3k-3)/(3k-2)$. Tightness of the bound is proved by the series of problems with $n = 2k$, $w_j = 2R$ for $j < k$, $w_k = R+1$, $w_j = R$ for $j > k$ and $c = (3k-2)R$ (e.g., for $k = 4$, $(w_j) = (2R, 2R, 2R, R+1, R, R, R, R)$, $c = 10R$). The unique optimal solution, of value $z = (3k-2)R$, includes all the items but the kth. Performing MTSS(k), there is no iteration in which M contains all items $j < k$, so the optimal solution could be found only by a greedy search starting from an item $j < k$. All such searches, however, will certainly include item k (since, at each iteration, at least two items of weight R are not in M), hence producing a solution value not greater than the greedy solution value $z^h = z^g = (3k-3)R + 1$. It follows that the ratio z^h/z can be arbitrarily close to $(3k-3)/(3k-2)$ for R sufficiently large. □

MTSS(k) dominates the Johnson (1974) scheme J(k), in the sense that, for any $k > 1$, the time complexity of both schemes is $O(n^k)$, while $r(\text{MTSS}(k)) = (3k-3)/(3k-2) > k/(k+1) = r(\text{J}(k))$ (for example: $r(\text{MTSS}(2)) = \frac{3}{4} = r(\text{J}(3))$, $r(\text{MTSS}(3)) = \frac{6}{7} = r(\text{J}(6))$, $r(\text{MTSS}(4)) = \frac{9}{10} = r(\text{J}(9))$). Also note that, for increasing values of k, the solution values returned by MTSS(k) are non-decreasing (because of the definition of M), while those returned by J(k) are not (if, for example, $(w_j) = (8, 5, 5, 3)$ and $c = 12$, J(1) returns $z^h = 11$, while J(2) returns $z^h = 10$).

Example 4.2 (continued)

We have already seen that MTSS(2) gives, in $O(n^2)$ time: $z^h = 96$, $X^h = \{4, 5, 6\}$. MTSS(3) gives, in $O(n^3)$ time: $z^h = 100$, $X^h = \{2, 8, 9\}$ (optimal). The solution is found when $M = \{2\}$ and the greedy search is performed starting from item 8. □

A more effective implementation of MTSS(k) can be obtained if, at each iteration i in the execution of MTGS, we update a pair $(L, \tilde{c})$ having the property that all items in $B = \{i, \dots, n\} \backslash (M \cup L)$ will be selected by the greedy search starting from i, and $\tilde{c} = c - \sum_{j \in B} w_j$. In this way, the greedy search can be performed only for the items in L with residual capacity $\tilde{c}$. Since each iteration removes items from L, execution of MTGS can be halted as soon as $L = \emptyset$. The improved version of MTSS(k) is obtained by replacing the call to MTGS with the statement

call MTGSM,

where:

procedure MTGSM:
input: $n, c, (w_j), M, g, z^h$;
output: z^g, V;
begin
    $z^g := z^h$;
    $L := \{1, \dots, n\} \backslash M$;
    $\tilde{c} := c - g$;
    $S := \emptyset$;
    $i := 0$;
    repeat
        $i := i + 1$;
        if $i \notin M$ **then**
            begin
                while $L \neq \emptyset$ and $w_j \leq \tilde{c}$ (j the first item in L) **do**
                    begin
                        $\tilde{c} := \tilde{c} - w_j$;
                        $S := S \cup \{j\}$;
                        $L := L \backslash \{j\}$
                    end;
                $\hat{c} := \tilde{c}$;
                $T := S$;
                for each $j \in L$ **do if** $w_j \leq \hat{c}$ **then**
                    begin
                        $\hat{c} := \hat{c} - w_j$;
                        $T := T \cup \{j\}$
                    end;

```
            if c − ĉ > z^g then
                begin
                    z^g := c − ĉ;
                    V := T
                end;
            c̃ := c̃ + w_i;
            S := S\{i}
        end;
    until L = Ø or z^g = c
end.
```

Example 4.2 (continued)

Calling MTGSM with $z^h = 0$, $M = \emptyset$ and $g = 0$, the execution is:

$L = \{1, \dots, 9\}$, $\tilde{c} = 100$, $S = \emptyset$;

$i = 1$: $\tilde{c} = 19$, $S = \{1\}$, $L = \{2, \dots, 9\}$;
$\hat{c} = 7$, $T = \{1, 7\}$, $z^g = 93$, $V = \{1, 7\}$;
$\tilde{c} = 100$, $S = \emptyset$;

$i = 2$: $\tilde{c} = 20$, $S = \{2\}$, $L = \{3, \dots, 9\}$;
$\hat{c} = 8$, $T = \{2, 7\}$;
$\tilde{c} = 100$, $S = \emptyset$;

$i = 3$: $\tilde{c} = 17$, $S = \{3, 4\}$, $L = \{5, \dots, 9\}$;
$\hat{c} = 5$, $T = \{3, 4, 7\}$, $z^g = 95$, $V = \{3, 4, 7\}$;
$\tilde{c} = 60$, $S = \{4\}$;

$i = 4$: $\tilde{c} = 4$, $S = \{4, 5, 6\}$, $L = \{7, 8, 9\}$;
$\hat{c} = 4$, $T = \{4, 5, 6\}$, $z^g = 96$, $V = \{4, 5, 6\}$;
$\tilde{c} = 44$, $S = \{5, 6\}$;

$i = 5$: $\tilde{c} = 12$, $S = \{5, 6, 7, 8, 9\}$, $L = \emptyset$;
$\hat{c} = 12$, $T = \{5, 6, 7, 8, 9\}$;
$\tilde{c} = 42$, $S = \{6, 7, 8, 9\}$. □

For large values of n, the computing time required by MTSS(k) can be further reduced in much the same way used for MTSL (Section 4.2.3), i.e. by determining the solution for an approximate core problem and then checking whether the requested performance (evaluated with respect to upper bound c on z) has been obtained.

Fischetti (1989) has proposed a polynomial-time approximation scheme, FS(k), based on the subdivision of N into a set of "small" items and a number of sets of "large" items, each containing items of "almost equal" weight. Although the worst-case performance ratio of the scheme has not been determined, it has been proved that $r(\text{FS}(k)) \geq ((k+2)^2 - 4)/(k+2)^2$. With this ratio, the result is $r(\text{MTSS}(k)) > r(\text{FS}(k))$ for $k < 6$, while $r(\text{MTSS}(k)) < r(\text{FS}(k))$ for $k > 6$.

4.3.3 Fully polynomial-time approximation schemes

The algorithms of the previous section allow one to obtain any prefixed worst-case performance ratio r in polynomial time and with linear space. The time complexity, however, is exponential in the inverse of the worst-case relative error $\varepsilon = 1 - r$.

The fully polynomial-time approximation scheme proposed by Ibarra and Kim (1975) for the 0-1 knapsack problem (procedure IK(ε) of Section 2.8.2) also applies to SSP. No sorting being required, the time complexity decreases from $O(n\log n) + O(n/\varepsilon^2)$ to $O(n/\varepsilon^2)$, polynomial in $1/\varepsilon$, while the space complexity remains $O(n + (1/\varepsilon^3))$. Lawler (1979) adapted to SSP his improved version of the Ibarra and Kim (1975) scheme for the 0-1 knapsack problem, obtaining time complexity $O(n + (1/\varepsilon^3))$ and space complexity $O(n + (1/\varepsilon^2))$, or time and space complexity $O(n + (1/\varepsilon^2)\log(1/\varepsilon))$.

All of the above schemes are based on the same idea, i.e. (see Section 2.8.2): (a) partitioning the items, basing on the value of ε, into "large" and "small" ones; (b) solving the problems for the large items only, with scaled weights, through dynamic programming; (c) completing the solution, in a greedy way, with the small items. Gens and Levner (1978, 1980) have proposed a fully polynomial-time approximation scheme based on a different (and simpler) principle. They solve the complete problem through ordinary dynamic programming but, at each iteration, reduce the current dynamic programming lists by keeping only state weights differing from each other by at least a threshold value depending on ε. The scheme can be conveniently defined using procedure RECS of Section 4.2.1. Note that the algorithm results similar to procedure DPS for the exact dynamic programming solution of SSP (Section 4.2.1). The main difference consists in determining, after each RECS call, reduced lists $W1$ and $X1$, instead of simply renaming $W2$ and $X2$ as $W1$ and $X1$.

procedure GL(ε):
input: $n, c, (w_j)$;
output: z^h, X^h;
begin
  determine $\sigma = \max\{j : \sum_{i=1}^{j} w_i \leq c\}$;
  $\tilde{z} := \max(\sum_{j=1}^{\sigma} w_j, \max_j\{w_j\})$ (**comment**: $\tilde{z} \leq z < 2\tilde{z}$);
  $\tilde{c} := c$;
  $W1_0 := 0$;
  $X1_0 := 0$;
  $s := 1$;
  $b := 2$;
  $W1_1 := w_1$;
  $X1_1 := 1$;
  $m := 2$;
  repeat

```
call RECS ;
h := 0;
j := 0;
repeat
    if W2_{j+1} > W1_h + εz̃ then j := j + 1
    else j := max{q : W2_q ≤ W1_h + εz̃};
    h := h + 1;
    W1_h := W2_j;
    X1_h := X2_j
until j = s;
m := m + 1;
s := h
until m > n or W1_s = c;
z^h := W1_s;
determine X^h by decoding X1_s
end.
```

At each iteration, the reduced dynamic programming lists clearly satisfy $W1_{h+2} - W1_h > \varepsilon\tilde{z}$ for $h = 1, \ldots, s-2$. Hence the number of states is always bounded by $s < 2z/(\varepsilon\tilde{z})$, that is, from $z < 2\tilde{z}$, by $s < (4/\varepsilon)$. It follows that the scheme has time and space complexity $O(n/\varepsilon)$. The proof that the solution determined by GL(ε) has worst-case relative error not greater than ε is given in Levner and Gens (1978) and Gens and Levner (1978).

The time and space complexity of GL(ε) can be better or worse than that of the Lawler (1979) scheme, according to the values of n and ε.

Example 4.2 (continued)

Calling GL($\frac{1}{3}$) , we initially find $\sigma = 1$, $\tilde{z} = 81$ and the weight list $W1$ given in the first column of Figure 4.4. No state is eliminated for $m = 2, 3$. For $m = 4$, $W2_4$ is eliminated since $W2_5 - W2_3 = 3 \leq \varepsilon\tilde{z} = 27$. The approximate solution found has the final value of $W1_4$, i.e. $z^h = 96$ (with $X^h = \{4, 5, 6\}$). □

4.3.4 Probabilistic analysis

As for the 0-1 knapsack problem (Section 2.8.3), we give a brief outline of the main results obtained in probabilistic analysis.

The first probabilistic result for SSP was obtained by d'Atri and Puech (1982). Assuming that the weights are independently drawn from a uniform distribution over $\{1, 2, \ldots, c(n)\}$ and the capacity from a uniform distribution over $\{1, 2, \ldots, nc(n)\}$, where $c(n)$ is an upper bound on the weights value, they proved that a simple variant of the greedy algorithm solves SSP with probability tending to 1.

Lagarias and Odlyzko (1983) considered SSP with equality constraint and assumed that the weights are independently drawn from a uniform distribution

	$m = 1$	$m = 2$		$m = 3$		$m = 4$		$m = 5$		$m = 6$		$m = 7$		$m = 8$		$m = 9$	
h	$W1$	$W2$	$W1$	$W2$	$W1$	$W2$	$W1$	$W2$	$W1$	$W2$	$W1$	$W2$	$W1$	$W2$	$W1$	$W2$	$W1$
0	0	0	0	0	0	0	0	0	0	0	0	0	0	0	0	0	0
1	81	80	80	43	43	40	40	30	30	26	26	12	26	11	26	9	26
2		81	81	80	80	43	43	40	43	30	43	26	43	26	43	26	52
3				81	81	80	80	43	70	43	70	38	70	37	70	35	79
4						81	83	70	83	56	96	43	96	43	96	43	96
5						83		73		69		55		54		52	
6								80		70		70		70		70	
7								83		83		82		81		79	
8										96		96		96		96	

Figure 4.4 State weights produced by $GL(\frac{1}{3})$ for Example 4.2

over $\{1, 2, \ldots, 2^{cn^2}\}$ and the capacity is the total weight of a randomly chosen subset of the items. They presented a polynomial-time algorithm which finds the solution for "almost all" instances with $c > 1$. The result was extended to $c > \frac{1}{2}$ by Frieze (1986).

The probabilistic properties of a "bounded" variant of SSP were investigated by Tinhofer and Schreck (1986).

4.4 COMPUTATIONAL EXPERIMENTS

In the present section we analyse the experimental behaviour of exact and approximate algorithms for SSP on random and deterministic test problems.

The main class of randomly generated test problems we use is

(i) *problems* $P(E)$: w_j uniformly random in $[1, 10^E]$;

$$c = n\,\frac{10^E}{4}.$$

For each pair (n, E), the value of c is such that about half the items can be expected to be in the optimal solution. In all algorithms for SSP, execution is halted as soon as a solution of value c is found. Hence the difficulty of a problem instance is related to the number of different solutions of value c. It follows that problems $P(E)$ tend to be more difficult when E grows. As we will see, truly difficult problems can be obtained only with very high values of 10^E. This confirms, in a sense, a theoretical result obtained by Chvátal (1980), who proved that, for the overwhelming majority of problems $P(n/2)$ (with n large enough), the running time of any algorithm based on branch-and-bound and dynamic programming is proportional at least to $2^{n/10}$. The Chvátal problems, as well as problems $P(E)$ with very high values of E, cannot be generated in practice because of the integer overflow limitation. A class of difficult problems which does not have this drawback is

(ii) *problems EVEN/ODD* : w_j even, uniformly random in $[1, 10^3]$;

$$c = \frac{n\,10^3}{4} + 1 \quad \text{(odd)}.$$

Since these problems admit no solution of value c, the execution of any enumerative algorithm terminates only after complete exploration of the branch-decision tree. Deterministic problems with the same property have been found by Todd (1980) and Avis (1980):

(iii) *problems TODD* : $w_j = 2^{k+n+1} + 2^{k+j} + 1$, with $k = \lfloor \log_2 n \rfloor$;

$$c = \left\lfloor 0.5 \sum_{j=1}^{n} w_j \right\rfloor = (n+1)2^{k+n} - 2^k + \left\lfloor \frac{n}{2} \right\rfloor,$$

(iv) *problems AVIS* : $w_j = n(n+1) + j;$

$$c = \left\lfloor \frac{n-1}{2} \right\rfloor n(n+1) + \binom{n}{2}.$$

4.4.1 Exact algorithms

We first compare, on small-size difficult problems, the Fortran IV implementations of the dynamic programming algorithm of Ahrens and Finke (1975) and of algorithm MTSL (Section 4.2.3). We used a CDC-Cyber 730 computer, having 48 bits available for integer operations, in order to be able to work with the large coefficients generated.

Table 4.1 gives the results for problems $P(E)$, with $E = 3, 6, 12$, Table 4.2 those for problems *EVEN/ODD, TODD, AVIS*. Each entry gives the average running

Table 4.1 Problems $P(E)$. CDC-Cyber 730 in seconds. Average times over 10 problems

	$P(3)$: w_j uniformly random in [1, 10^3]; $c = n10^3/4$		$P(6)$: w_j uniformly random in [1, 10^6]; $c = n10^6/4$		$P(12)$: w_j uniformly random in [1, 10^{12}]; $c = n10^{12}/4$	
n	Ahrens and Finke	MTSL	Ahrens and Finke	MTSL	Ahrens and Finke	MTSL
8	0.012	0.004	0.012	0.004	0.013	0.004
12	0.023	0.010	0.029	0.013	0.029	0.013
16	0.040	0.011	0.091	0.049	0.092	0.050
20	0.069	0.007	0.322	0.185	0.422	0.232
24	0.137	0.010	0.640	0.513	2.070	1.098
28	0.349	0.010	1.341	0.647	9.442	6.306
32	0.940	0.009	2.284	0.661	time limit	time limit
36	2.341	0.009	4.268	0.605	—	—
40	5.590	0.011	9.712	0.663	—	—

Table 4.2 Problems *EVEN/ODD*, *TODD*, *AVIS*. CDC-Cyber 730 in seconds

	EVEN/ODD; average times over 10 problems		*TODD*; single trials		*AVIS*; single trials	
n	Ahrens and Finke	MTSL	Ahrens and Finke	MTSL	Ahrens and Finke	MTSL
8	0.013	0.005	0.013	0.002	0.016	0.002
12	0.028	0.021	0.050	0.005	0.041	0.005
16	0.090	0.053	0.199	0.020	0.111	0.012
20	0.392	0.190	0.785	0.257	0.326	0.046
24	1.804	0.525	3.549	0.400	0.815	0.126
28	7.091	0.969	15.741	0.403	2.010	0.291
32	21.961	1.496	70.677	0.407	4.348	0.579
36	time limit	2.184	308.871	0.409	8.345	1.146
40	—	2.941	—	—	15.385	1.780

time, expressed in seconds, computed over 10 problem instances (except for the deterministic *TODD* and *AVIS* problems, for which single runs were executed). Each algorithm had a time limit of 450 seconds to solve the problems generated for each data set. MTSL was always faster than the Ahrens and Finke (1975) algorithm. Table 4.1 shows that problems $P(E)$ become really hard only when very high values of 10^E are employed. Table 4.2 demonstrates that the "artificial" hard problems can still be solved, in reasonable time, by MTSL. (Problems *TODD* cannot be generated for $n \geq 40$ because of integer overflows.)

We used a 32-bit HP 9000/840 computer, having a core memory limitation of 10 Mbytes, to test MTSL on very large "easy" $P(E)$ instances. Since the Fortran implementation of MTSL requires only two vectors of dimension n, we were able to solve problems up to one million variables. Because of integer overflow limitations, the capacity was set to $n10^E/50$, hence E could not be greater than 5. Table 4.3 gives the average times, computed over 20 problem instances, relative to problems $P(2)$, $P(3)$, $P(4)$, $P(5)$. The results show very regular times, growing almost linearly with n. No remarkable difference comes from the different values of E used. The initial value of ϑ ($\vartheta = 45$) always produced the optimal solution. All runs were executed with option "-o" for the Fortran compiler, i.e. with the lowest optimization level.

Table 4.3 Problems $P(E)$. HP 9000/840 in seconds. Average times over 20 problems

n	$P(2)$ w_j uniformly random in $[1, 10^2]$; $c = n10^2/50$	$P(3)$ w_j uniformly random in $[1, 10^3]$; $c = n10^3/50$	$P(4)$ w_j uniformly random in $[1, 10^4]$; $c = n10^4/50$	$P(5)$ w_j uniformly random in $[1, 10^5]$; $c = n10^5/50$
1 000	0.007	0.010	0.022	0.125
2 500	0.009	0.014	0.025	0.116
5 000	0.016	0.020	0.031	0.121
10 000	0.028	0.032	0.046	0.126
25 000	0.070	0.071	0.088	0.173
50 000	0.136	0.138	0.156	0.252
100 000	0.277	0.272	0.295	0.392
250 000	0.691	0.674	0.716	0.801
500 000	1.361	1.360	1.418	1.527
1 000 000	2.696	2.720	2.857	2.948

4.4.2 Approximate algorithms

We used the hard problems of the previous section to experimentally compare approximate algorithms for SSP. The runs were executed on a CDC-Cyber 730 computer, with values of n ranging from 10 to 1 000 for problems *EVEN/ODD* and $P(10)$ ($E = 10$ being the maximum value not producing integer overflows), from 10 to 35 for problems *TODD*. We compared the Fortran IV implementations of the polynomial-time approximation schemes of Johnson (1974) and Martello and Toth (1984b) and those of the fully polynomial-time approximation schemes of Lawler (1979) and Gens and Levner (1978, 1980) (referred to as J(k), MTSS(k), L(ε) and GL(ε), respectively). The size of the approximate core for MTSS(k) was set to $200/k$.

We used the values $\frac{1}{2}$, $\frac{3}{4}$ and $\frac{6}{7}$ of the worst-case performance ratio r. These are the smallest values which can be imposed on all the schemes. Table 4.4 shows the parameters used and the time and space complexities.

Table 4.4 Time and space complexities

	J(k)			MTSS(k)			L(ε)			GL(ε)	
r	k	time	space	k	time	space	ε	time	space	time	space
$\frac{1}{2}$	1	$O(n)$	$O(n)$	1	$O(n)$	$O(n)$	$\frac{1}{2}$	$O(n+\frac{1}{\varepsilon^3})$	$O(n+\frac{1}{\varepsilon^2})$	$O(\frac{n}{\varepsilon})$	$O(\frac{n}{\varepsilon})$
$\frac{3}{4}$	3	$O(n^3)$	$O(n)$	2	$O(n^2)$	$O(n)$	$\frac{1}{4}$	$O(n+\frac{1}{\varepsilon^3})$	$O(n+\frac{1}{\varepsilon^2})$	$O(\frac{n}{\varepsilon})$	$O(\frac{n}{\varepsilon})$
$\frac{6}{7}$	6	$O(n^6)$	$O(n)$	3	$O(n^3)$	$O(n)$	$\frac{1}{7}$	$O(n+\frac{1}{\varepsilon^3})$	$O(n+\frac{1}{\varepsilon^2})$	$O(\frac{n}{\varepsilon})$	$O(\frac{n}{\varepsilon})$

For each triple (type of problem, value of r, value of n), we generated ten problems and solved them with the four algorithms. The tables give two types of entries: average running times and average percentage errors. The errors were computed with respect to the optimal solution for problems *TODD*. For problems $P(10)$ and *EVEN/ODD* we used the optimal solution when $n < 50$, and the upper bound c (for $P(10)$) or $c-1$ (for *EVEN/ODD*) when $n \geq 50$. When all ten problems were exactly solved by an algorithm, the corresponding error entry is "exact" (entry 0.0000 means that the average percentage error was less than 0.00005).

Table 4.5 gives the results for problems $P(10)$. L(ε) has, in general, very short times and very large errors. This is because the number of large items is very small (for $n \leq 50$) or zero (for $n \geq 100$). MTSS(k) dominates the other algorithms, J(k) dominates GL(ε). For any $n \geq 25$, J(k) gives exactly the same results, independently of r since, for all such cases, set L is empty, so only the greedy algorithm is performed. The running times of GL(ε) grow with n and with r, those of J(k) only with n, those of MTSS(k) only with r (for $n \geq 50$), while L(ε) has an irregular behaviour.

Table 4.6 gives the results for problems *EVEN/ODD*. As in Table 4.5, L(ε) always has very short times and very large errors, MTSS(k) dominates the other algorithms and J(k) dominates GL(ε). The running times and the growing rates of errors are the same as in Table 4.5 while the absolute errors are different. In many cases MTSS(k) found the optimal solution; since, however, the corresponding value does not coincide with c, execution could not stop, so the running times grow with r.

Table 4.7 gives results for problems *TODD*. Since these problems are deterministic, the entries refer to single trials. MTSS(k) dominates all the

Table 4.5 Problems $P(10)$: w_j uniformly random in $[1, 10^{10}]$; $c = n10^{10}/4$. CDC-Cyber 730 in seconds. Average values over 10 problems

		Time				Percentage error			
n	r	J(k)	MTSS(k)	L(ε)	GL(ε)	J(k)	MTSS(k)	L(ε)	GL(ε)
	$\frac{1}{2}$	0.001	0.001	0.004	0.005	2.0871	2.0871	5.5900	2.0307
10	$\frac{3}{4}$	0.001	0.001	0.012	0.009	2.0044	0.4768	3.7928	1.2864
	$\frac{6}{7}$	0.003	0.006	0.025	0.014	0.8909	0.1894	2.8857	0.9088
	$\frac{1}{2}$	0.002	0.003	0.001	0.014	0.3515	0.3515	5.3916	1.8044
25	$\frac{3}{4}$	0.002	0.005	0.008	0.020	0.3515	0.0467	1.9958	0.6695
	$\frac{6}{7}$	0.003	0.035	0.069	0.037	0.3515	0.0049	1.5973	0.6100
	$\frac{1}{2}$	0.004	0.005	0.001	0.029	0.0833	0.0833	0.8870	0.2519
50	$\frac{3}{4}$	0.004	0.009	0.001	0.050	0.0833	0.0058	0.8870	0.1008
	$\frac{6}{7}$	0.004	0.166	0.016	0.079	0.0833	0.0002	0.9902	0.0794
	$\frac{1}{2}$	0.009	0.014	0.002	0.061	0.0082	0.0082	1.0991	0.0611
100	$\frac{3}{4}$	0.009	0.020	0.001	0.093	0.0082	0.0004	1.0991	0.0708
	$\frac{6}{7}$	0.010	0.207	0.001	0.157	0.0082	0.0001	1.0991	0.0541
	$\frac{1}{2}$	0.020	0.022	0.003	0.112	0.0032	0.0039	0.7441	0.0077
250	$\frac{3}{4}$	0.022	0.029	0.003	0.235	0.0032	0.0004	0.7441	0.0070
	$\frac{6}{7}$	0.022	0.158	0.003	0.374	0.0032	0.0000	0.7441	0.0059
	$\frac{1}{2}$	0.049	0.022	0.008	0.254	0.0010	0.0040	0.2890	0.0016
500	$\frac{3}{4}$	0.042	0.033	0.006	0.438	0.0010	0.0001	0.2890	0.0016
	$\frac{6}{7}$	0.047	0.180	0.007	0.685	0.0010	0.0000	0.2890	0.0015
	$\frac{1}{2}$	0.100	0.024	0.013	0.540	0.0002	0.0014	0.1954	0.0005
1000	$\frac{3}{4}$	0.102	0.030	0.013	0.909	0.0002	0.0001	0.1954	0.0006
	$\frac{6}{7}$	0.102	0.224	0.013	1.374	0.0002	0.0000	0.1954	0.0005

Table 4.6 Problems *EVEN/ODD*. CDC-Cyber 730 in seconds. Average times over 10 problems

		Time				Percentage error			
n	r	J(k)	MTSS(k)	L(ε)	GL(ε)	J(k)	MTSS(k)	L(ε)	GL(ε)
	$\frac{1}{2}$	0.001	0.001	0.005	0.005	2.2649	2.2649	7.5859	1.5131
10	$\frac{3}{4}$	0.001	0.002	0.012	0.009	2.3209	0.8325	3.1369	0.8403
	$\frac{6}{7}$	0.003	0.007	0.025	0.013	0.9202	0.0720	2.5209	0.9041
	$\frac{1}{2}$	0.002	0.003	0.001	0.015	0.2432	0.2432	7.6416	1.0688
25	$\frac{3}{4}$	0.002	0.005	0.011	0.026	0.2432	0.0384	2.3360	0.4288
	$\frac{6}{7}$	0.002	0.048	0.077	0.042	0.2432	exact	1.9808	0.3584
	$\frac{1}{2}$	0.004	0.006	0.001	0.030	0.0400	0.0400	2.4480	0.1424
50	$\frac{3}{4}$	0.004	0.011	0.001	0.051	0.0400	0.0016	2.4480	0.1680
	$\frac{6}{7}$	0.004	0.287	0.019	0.084	0.0400	exact	1.1232	0.0816
	$\frac{1}{2}$	0.009	0.011	0.002	0.060	0.0160	0.0160	0.7352	0.0792
100	$\frac{3}{4}$	0.009	0.028	0.002	0.100	0.0160	exact	0.7352	0.0792
	$\frac{6}{7}$	0.008	0.274	0.002	0.166	0.0160	exact	0.7352	0.0520
	$\frac{1}{2}$	0.022	0.022	0.003	0.141	0.0019	0.0006	0.4221	0.0080
250	$\frac{3}{4}$	0.021	0.028	0.004	0.235	0.0019	exact	0.4221	0.0070
	$\frac{6}{7}$	0.021	0.242	0.003	0.380	0.0019	exact	0.4221	0.0058
	$\frac{1}{2}$	0.047	0.026	0.007	0.291	0.0003	0.0006	0.2682	0.0021
500	$\frac{3}{4}$	0.047	0.031	0.007	0.483	0.0003	exact	0.2682	0.0021
	$\frac{6}{7}$	0.047	0.257	0.007	0.774	0.0003	exact	0.2682	0.0024
	$\frac{1}{2}$	0.104	0.029	0.013	0.595	exact	0.0010	0.1325	0.0002
1000	$\frac{3}{4}$	0.104	0.033	0.014	0.992	exact	exact	0.1325	0.0002
	$\frac{6}{7}$	0.104	0.293	0.014	1.567	exact	exact	0.1325	0.0001

Table 4.7 Problems *TODD*. CDC-Cyber 730 in seconds. Single trials

		Time				Percentage error			
n	r	J(k)	MTSS(k)	L(ε)	GL(ε)	J(k)	MTSS(k)	L(ε)	GL(ε)
	$\frac{1}{2}$	0.001	0.001	0.002	0.005	9.9721	9.9721	8.2795	exact
10	$\frac{3}{4}$	0.001	0.002	0.013	0.008	9.9721	exact	4.4157	exact
	$\frac{6}{7}$	0.004	0.006	0.022	0.011	exact	exact	2.1366	exact
	$\frac{1}{2}$	0.001	0.001	0.001	0.008	exact	exact	12.3660	6.1343
15	$\frac{3}{4}$	0.002	0.002	0.016	0.016	exact	exact	6.2072	3.0185
	$\frac{6}{7}$	0.001	0.012	0.033	0.023	exact	exact	1.4548	exact
	$\frac{1}{2}$	0.001	0.003	0.001	0.014	4.7761	4.7761	4.7482	4.7482
20	$\frac{3}{4}$	0.001	0.003	0.006	0.023	4.7761	exact	4.7482	exact
	$\frac{6}{7}$	0.002	0.018	0.036	0.040	4.7761	exact	2.3787	exact
	$\frac{1}{2}$	0.001	0.003	0.001	0.016	exact	exact	7.6896	7.6896
25	$\frac{3}{4}$	0.001	0.004	0.001	0.036	exact	exact	7.6896	exact
	$\frac{6}{7}$	0.001	0.028	0.048	0.059	exact	exact	3.3638	1.9210
	$\frac{1}{2}$	0.002	0.003	0.001	0.017	3.2261	3.2261	3.2255	3.2255
30	$\frac{3}{4}$	0.003	0.005	0.001	0.030	3.2261	exact	3.2255	3.2253
	$\frac{6}{7}$	0.002	0.038	0.017	0.052	3.2261	exact	0.8063	exact
	$\frac{1}{2}$	0.003	0.004	0.001	0.021	exact	exact	5.5555	2.7777
35	$\frac{3}{4}$	0.002	0.004	0.001	0.035	exact	exact	5.5555	2.7777
	$\frac{6}{7}$	0.002	0.045	0.015	0.062	exact	exact	1.3888	exact

algorithms, while L(ε) is generally dominated by all the algorithms. J(k) dominates GL(ε) for n odd (J(1) always finds the optimal solution). For n even, GL(ε) has higher times but much smaller errors than J(k), MTSS(2) always finds the optimal solutions, MTSS(1) only for n odd. This behaviour of the algorithms can be explained by analysing the structure of the optimal solution to problems *TODD*. Let $m = \lfloor n/2 \rfloor$, so $c = (n+1)2^{k+n} - 2^k + m$. Hence the number of items in any feasible solution is at most m since, by algebraic manipulation, the sum of the $m + 1$ smallest weights is

$$\sum_{i=1}^{m+1} w_i = 2(m+1)2^{k+n} + 2^{k+1}(2^{m+1} - 1) + (m+1) > c$$

(in problems *TODD* the w_i's are given for increasing values). For n odd ($n = 2m + 1$), the sum of the m largest weights is feasible, since

$$\sum_{i=n-m+1}^{n} w_i = (2m+2)2^{k+n} - 2^{k+n-m+1} + m < c,$$

and hence optimal. So, after sorting, the greedy algorithm (J(1) or MTSS(1)) finds the optimal solution. For n even ($n = 2m$), (a) any solution including w_n includes at most $m - 2$ further items, since

$$w_n + \sum_{i=1}^{m-1} w_i = (2m+1)2^{k+n} + 2^k(2^m - 2) + m > c;$$

(b) it follows that the best solution including w_n has value

$$z_1 = \sum_{i=n-m+2}^{n} w_i = 2m2^{k+n} - 2^{k+n-m+2} + m - 1 < c;$$

(c) the best solution not including w_n has value

$$z_2 = \sum_{i=n-m}^{n-1} w_i = (2m+1)2^{k+n} - 2^{k+n-m} + m < c,$$

and $z_2 > z_1$. So z_2 is the optimal solution value and MTSS(2) finds it when, after sorting, it applies the greedy algorithm starting from the second element.

We do not give the results for problems *AVIS*, for which the algorithms have a behaviour similar to that of problems *TODD*. In fact, let $s = \lfloor (n-1)/2 \rfloor$, so $c = sn(n+1) + n(n-1)/2$. Since the sum of the $s + 1$ smallest weights is

$$\sum_{i=1}^{s+1} w_i = sn(n+1) + n(n+1) + \tfrac{1}{2}\,(s+1)(s+2) > c,$$

any feasible solution will include, at most, s items. The sum of the s largest weights is feasible, since

$$\sum_{i=n-s+1}^{n} w_i = sn(n+1) + s(n-s) + \tfrac{1}{2}\,s(s+1) \leq c,$$

hence optimal. So, after sorting, the greedy algorithms $J(1)$ and MTSS(1) always find the optimal solution.

The computational results of this section (and others, reported in Martello and Toth (1985a)) show that all the polynomial-time approximation schemes for SSP have an average performance much better than their worst-case performance. So, in practical applications, we can obtain good results with short computing times, i.e. by imposing small values of the worst-case performance ratio.

Although polynomial-time approximation schemes have a worse bound on computing time, their average performance appears superior to that of the fully polynomial-time approximation schemes, in the sense that they generally give better results with shorter computing times and fewer core memory requirements.

The most efficient scheme is MTSS(k). For $n \geq 50$, the largest average error of MTSS(2) was 0.0075 per cent, that of MTSS(3) 0.0005 per cent. So, for practical purposes, one of these two algorithms should be selected while using higher values of k would probably be useless.

5

Change-making problem

5.1 INTRODUCTION

In Chapter 1 the change-making problem has been presented, for the sake of uniformity, as a maximization problem with bounded variables. However, in the literature it is generally considered in minimization form and, furthermore, the main results have been obtained for the case in which the variables are unbounded. Hence we treat the bounded case in the final section of this chapter, the remaining ones being devoted to the *Change-Making Problem* (CMP) defined as follows. Given n *item types* and a *knapsack*, with

$$w_j = \textit{weight} \text{ of an item of type } j;$$

$$c = \textit{capacity} \text{ of the knapsack,}$$

select a number x_j $(j = 1, \dots, n)$ of items of each type so that the total weight is c and the total number of items is a minimum, i.e.

$$\text{minimize} \quad z = \sum_{j=1}^{n} x_j \tag{5.1}$$

$$\text{subject to} \quad \sum_{j=1}^{n} w_j x_j = c, \tag{5.2}$$

$$x_j \geq 0 \text{ and integer}, \quad j \in N = \{1, \dots, n\}. \tag{5.3}$$

The problem is NP-hard also in this version, since Lueker (1975) has proved that even the feasibility question (5.2)–(5.3) is NP-complete. The problem is called "change-making" since it can be interpreted as that of a cashier having to assemble a given change, c, using the least number of coins of specified values w_j $(j = 1, \dots, n)$ in the case where, for each value, an unlimited number of coins is available. CMP can also be viewed as an unbounded knapsack problem (Section 3.6) in which $p_j = -1$ for all j and, in the capacity constraint, strict equality is imposed. (On the other hand, imposing inequality $\sum_{j=1}^{n} w_j x_j \geq c$ gives rise to a trivial problem whose optimal solution is $x_l = \lceil c/w_l \rceil$ (where l is the item type of maximum weight) and $x_j = 0$ for $j \in N \backslash \{l\}$, since item type l "dominates" all the others in the sense of Theorem 3.2.) Note that, because of (5.2), a feasible solution to the problem does not necessarily exist.

It is usual in the literature to consider positive weights w_j. Hence, we will also assume, without loss of generality, that

$$w_j \text{ and } c \text{ are integers;} \tag{5.4}$$

$$w_j < c \quad \text{for } j \in N; \tag{5.5}$$

$$w_i \neq w_j \quad \text{if } i \neq j. \tag{5.6}$$

Violation of assumption (5.4) can be handled by scaling. If assumption (5.5) is violated, then we can set $x_j = 0$ for all j such that $w_j > c$ and, if there is an item type (say k) with $w_k = c$, immediately determine an optimal solution ($x_k = 1$, $x_j = 0$ for $j \in N \backslash \{k\}$). If assumption (5.6) is violated then the two item types can be replaced by a single one. Note that, on the contrary, the assumption on the positivity of w_j ($j \in N$) produces a loss of generality, because of the equality constraint.

CMP can arise, in practice, in some classes of unidimensional cargo-loading and cutting stock problems. Consider, for example, a wall to be covered with panels: how is it possible, given the available panel lengths, to use the least possible number of panels?

In the following sections we examine lower bounds (Section 5.2), greedy solutions (Sections 5.3, 5.4), dynamic programming and branch-and-bound algorithms (Sections 5.5, 5.6), and the results of computational experiments (Section 5.7). Section 5.8 analyses the generalization of the problem to the case where, for each j, an upper bound on the availability of items of type j is given (*Bounded Change-Making Problem*).

5.2 LOWER BOUNDS

Assume, without loss of generality, that the item types satisfy

$$w_1 > w_2 > w_3 > w_j \quad \text{for} \quad j = 4, \dots, n. \tag{5.7}$$

Let us consider the continuous relaxation of CMP, i.e. (5.1), (5.2) and

$$x_j \geq 0, \qquad j \in N.$$

From (5.7), its optimal solution $\overline{x}$ is straightforward ($\overline{x}_1 = c/w_1, \overline{x}_j = 0$ for $j = 2, \dots, n$) and provides an immediate lower bound for CMP:

$$L_0 = \left\lceil \frac{c}{w_1} \right\rceil.$$

If we also impose, similarly to what has been done for the unbounded knapsack problem (Section 3.6.1), the obvious condition $\overline{x}_1 \leq \lfloor c/w_1 \rfloor$, which must hold in any integer solution, the continuous solution becomes

$$\overline{x}_1 = \left\lfloor \frac{c}{w_1} \right\rfloor,$$

$$\overline{x}_j = 0 \quad \text{for} \quad j = 3, \ldots, n,$$

$$\overline{x}_2 = \frac{\overline{c}}{w_2},$$

where

$$\overline{c} = c (\bmod w_1). \tag{5.8}$$

This gives an improved lower bound:

$$L_1 = \left\lfloor \frac{c}{w_1} \right\rfloor + \left\lceil \frac{\overline{c}}{w_2} \right\rceil.$$

Suppose now that $\lfloor c/w_1 \rfloor$ items of type 1 and $\lfloor \overline{c}/w_2 \rfloor$ items of type 2 are initially selected, and let

$$z' = \left\lfloor \frac{c}{w_1} \right\rfloor + \left\lfloor \frac{\overline{c}}{w_2} \right\rfloor, \tag{5.9}$$

$$c' = \overline{c}(\bmod w_2). \tag{5.10}$$

In the optimal solution to CMP, either $x_2 \leq \lfloor \overline{c}/w_2 \rfloor$ or $x_2 > \lfloor \overline{c}/w_2 \rfloor$. In the former case the continuous relaxation gives a lower bound

$$L^0 = z' + \left\lceil \frac{c'}{w_3} \right\rceil, \tag{5.11}$$

while in the latter a valid lower bound is

$$L^1 = z' - 1 + \left\lceil \frac{c' + w_1}{w_2} \right\rceil, \tag{5.12}$$

since the condition implies $x_1 \leq \lfloor c/w_1 \rfloor - 1$. We have thus proved the following

Theorem 5.1 (Martello and Toth, 1980b). *The value*

$$L_2 = \min (L^0, L^1),$$

where L^0 and L^1 are defined by (5.8)–(5.12), *is a lower bound for CMP.*

Since L_1 can be re-written as $z' + \lceil c'/w_2 \rceil$, we have $L^0 \geq L_1$. Also, by noting that $L^1 = z' + \lceil (c' + w_1 - w_2)/w_2 \rceil$ and $w_1 - w_2 > 0$, we have $L^1 \geq L_1$. Hence L_1 is dominated by the new bound.

The time complexity for the computation of the above bounds is clearly $O(n)$. No sorting is in fact needed, since only the three largest weights are required.

Example 5.1

Consider the instance of CMP defined by

$n = 5;$

$(w_j) = (11, 8, 5, 4, 1);$

$c = 29.$

We obtain

$L_0 = 3.$

$L_1 = 2 + \lceil \frac{7}{8} \rceil = 3.$

$L^0 = 2 + \lceil \frac{7}{5} \rceil = 4;$

$L^1 = 2 - 1 + \lceil \frac{18}{8} \rceil = 4;$

$L_2 = 4.$ □

Lower bounds L_0, L_1 and L_2 are the respective conceptual counterparts of upper bounds U_0, U_1 and U_3 introduced for the unbounded knapsack problem (Section 3.6.1), for which we have proved that the worst case performance ratio is 2. As often occurs, however, maximization and minimization problems do not have the same theoretical properties for upper and lower bounds (see, e.g., Section 2.4). For CMP, the worst-case performance of the lower bounds above is arbitrarily bad. Consider in fact the series of instances with $n = 3$, $w_1 = k$, $w_2 = k - 1$, $w_3 = 1$ and $c = 2k - 3$ $(k > 3)$: we have $L_0 = L_1 = L_2 = 2$, while the optimal solution has value $z = k - 2$, so the ratio L_i/z $(i = 1, 2 \text{ or } 3)$ can be arbitrarily close to 0 for k sufficiently large.

5.3 GREEDY ALGORITHMS

In the present section we consider both the change-making problem (5.1)–(5.3) and the generalization we obtain by associating an integer non-negative *cost* q_j with each item type $j \in N$ and changing the objective function to

$$\text{minimize} \quad z = \sum_{j=1}^{n} q_j x_j. \tag{5.13}$$

We obtain an *Unbounded Equality Constrained Min-Knapsack Problem* (UEMK). UEMK contains, as special cases:

(i) CMP, when $q_j = 1$ for all $j \in N$;

(ii) the unbounded knapsack problem (Section 3.6) in minimization form, when an extra item type $n+1$ is added, with $q_{n+1} = 0$ and $w_{n+1} = -1$.

For convenience of notation, we assume that the item types are sorted according to *decreasing* values of the cost per unit weight, i.e.

$$\frac{q_1}{w_1} \geq \frac{q_2}{w_2} \geq \ldots \geq \frac{q_n}{w_n}, \tag{5.14}$$

and note that, for CMP, this implies

$$w_1 < w_2 < \ldots < w_n.$$

A greedy algorithm for UEMK can be derived immediately from procedure GREEDYU of Section 3.6.1 as follows.

```
procedure GREEDYUM:
input: n, c, (q_j), (w_j);
output: z^g, (x_j), c̄;
begin
    c̄ := c;
    z^g := 0;
    for j := n to 1 step −1 do
        begin
            x_j := ⌊c̄/w_j⌋;
            c̄ := c̄ − w_j x_j;
            z^g := z^g + q_j x_j
        end
end.
```

The time complexity of GREEDYUM is $O(n)$, plus $O(n\log n)$ for the preliminary sorting. By replacing the last statement with

$$z^g := z^g + x_j$$

we obtain a greedy algorithm for CMP.

On return from GREEDYUM, if $\overline{c} = 0$ then z^g and (x_j) give a feasible (not necessarily optimal) solution. If $\overline{c} > 0$ then no feasible solution has been found by the procedure.

Example 5.1 (continued)

Applying GREEDYUM we obtain $(x_j) = (2, 0, 1, 0, 2)$, $z^g = 5$ and $\overline{c} = 0$, while the optimal solution has value $z = 4$ (as will be seen in Section 5.5.2). □

The case in which at least one item type has weight of value 1, has interesting properties. First, a feasible solution to the problem always exists. Furthermore, GREEDYUM always returns a feasible solution, since the iteration in which this item is considered produces $\overline{c} = 0$. The worst-case behaviour of the greedy algorithm, however, is arbitrarily bad, also with this restriction. Consider in fact the series of instances (both of CMP and UEMK) with: $n = 3$, $q_j = 1$ for all j, $w_1 = 1$, $w_2 = k$, $w_3 = k+1$ and $c = 2k > 2$, for which $z = 2$ and $z^g = k$, so the ratio z^g/z goes to infinity with k. (The absolute error produced by GREEDYUM for UEMK has been investigated by Magazine, Nemhauser and Trotter (1975), the relative error produced for CMP by Tien and Hu (1977).)

Consider now a change-making problem for the US coins, i.e. with: $n = 6$, $w_1 = 1, w_2 = 5$, $w_3 = 10$, $w_4 = 25$, $w_5 = 50$, $w_6 = 100$. It is not difficult to be convinced that GREEDYUM gives the optimal solution for any possible value of c (expressed in cents). Situations of this kind are analysed in the next section.

5.4 WHEN THE GREEDY ALGORITHM SOLVES CLASSES OF KNAPSACK PROBLEMS

We consider instances of CMP and UEMK in which at least one item type has weight of value 1.

For CMP, this implies that, after sorting,

$$1 = w_1 < w_2 < \ldots < w_n. \tag{5.15}$$

A weight vector $(w_1, \ldots, w_n)$ is called *canonical* if the greedy algorithm exactly solves all instances of CMP defined by the vector and any positive integer c. Chang and Gill have given the following necessary and sufficient condition for testing whether a vector is canonical.

Theorem 5.2 (Chang and Gill, 1970a) *A vector $(w_1, \ldots, w_n)$ satisfying (5.15) is canonical if and only if for all integers c in the range*

$$w_3 \le c < \frac{w_n(w_n w_{n-1} + w_n - 3w_{n-1})}{w_n - w_{n-1}}$$

the greedy solution is optimal.

The proof is quite involved and will not be reported here. Furthermore, application of the theorem is very onerous, calling for optimality testing of a usually high number of greedy solutions.

Example 5.2

Consider the instance of CMP defined by

$n = 7;$

$(w_j) = (1, 2, 4, 8, 10, 16).$

This vector can be proved to be canonical. However, application of the Chang and Gill (1970a) test requires us to solve, both in an exact and greedy way, 386 instances. □

We now turn to instances of UEMK. Let j^* denote an item type such that $w_{j^*} = 1$ and $q_{j^*} = \min\{q_j : w_j = 1\}$ and note that all item types k for which $q_k/w_k \geq q_{j^*}/w_{j^*}$ are "dominated" by j^* so they can be eliminated from the instance. Hence we assume, without loss of generality, that the item types, sorted according to (5.14), satisfy

$$1 = w_1 < w_j \quad \text{for } j = 2, \dots, n. \tag{5.16}$$

For $k = 1, \dots, n$ and for any positive integer c, let $f_k(c)$ and $g_k(c)$ denote, respectively, the optimal and greedy solution value to

$$\text{minimize} \quad \sum_{j=1}^{k} q_j x_j$$

$$\text{subject to} \quad \sum_{j=1}^{k} w_j x_j = c,$$

$$x_j \geq 0 \quad \text{and integer}, \quad j = 1, \dots, k.$$

When $f_k(c) = g_k(c)$ for all c—or, more concisely, $f_k = g_k$—we say that the pair of vectors $((q_1, \dots, q_k), (w_1, \dots, w_k))$ is *canonical*. The following theorem provides a recursive sufficient condition for checking whether a pair of vectors is so.

Theorem 5.3 (Magazine, Nemhauser and Trotter, 1975) *Assume that* $(q_1, \dots, q_n)$ *and* $(w_1, \dots, w_n)$ *satisfy* (5.14) *and* (5.16). *If, for fixed* k $(1 \leq k \leq n)$, $f_k = g_k$ *and* $w_{k+1} > w_k$, *then, by defining* $m = \lceil w_{k+1}/w_k \rceil$ *and* $\gamma = mw_k - w_{k+1}$, *the following are equivalent:*

(i) $f_{k+1} = g_{k+1}$,

(ii) $f_{k+1}(mw_k) = g_{k+1}(mw_k)$,

(iii) $q_{k+1} + g_k(\gamma) \leq mq_k$.

Proof. It is obvious that (ii) follows from (i). Since $g_{k+1}(mw_k) = q_{k+1} + g_k(\gamma)$ and $f_{k+1}(mw_k) \le mq_k$, (iii) follows from (ii). To prove that (i) follows from (iii), suppose, by absurdity, that (iii) holds but there exists a value c for which $f_{k+1}(c) < g_{k+1}(c)$. It must be $c > w_{k+1}$, since for $c < w_{k+1}$ we have $f_{k+1}(c) = f_k(c) = g_k(c) = g_{k+1}(c)$ while for $c = w_{k+1}$ we have $f_{k+1}(w_{k+1}) = g_{k+1}(w_{k+1}) = q_{k+1}$. Let $p = \lfloor c/w_k \rfloor$ and $\delta = c - pw_k$, and note that $p \ge m - 1$.

If $x_{k+1} > 0$ in an optimal solution, then $f_{k+1}(c) - g_{k+1}(c) = f_{k+1}(c - x_{k+1}w_{k+1}) - g_{k+1}(c - x_{k+1}w_{k+1})$ (since the greedy solution includes at least x_{k+1} items of type $k+1$), so we can assume that c is such that $x_{k+1} = 0$ in any optimal solution. Hence

$$f_{k+1}(c) = f_k(c) = g_k(c) = pq_k + f_k(\delta) = mq_k + (p - m)q_k + f_k(\delta). \tag{5.17}$$

From the definition of δ, by algebraic manipulation we can write $c = w_{k+1} + (p - m)w_k + \gamma + \delta$. Hence: (a) if $p \ge m$ then

$$f_{k+1}(c) < q_{k+1} + (p - m)q_k + g_k(\gamma) + f_k(\delta); \tag{5.18}$$

(b) if $p = m - 1$ then $\gamma + \delta - w_k = c - w_{k+1} > 0$ and $f_k(\gamma + \delta) = f_k(\gamma + \delta - w_k) + q_k$, so

$$f_{k+1}(c) < q_{k+1} + f_k(\gamma + \delta - w_k) \le q_{k+1} - q_k + g_k(\gamma) + f_k(\delta),$$

showing that (5.18) holds for all $p \ge m - 1$. Combining (5.17) and (5.18) we obtain $mq_k < q_{k+1} + g_k(\gamma)$, a contradiction. □ (An alternative proof has been given by Hu and Lenard (1976).)

Theorem 5.3 is known as the "one-point theorem" since, given a canonical pair $((q_1, \dots, q_k), (w_1, \dots, w_k))$ and a further item, $k + 1$, satisfying $q_{k+1}/w_{k+1} \ge q_k/w_k$ and $w_{k+1} > w_k$, the canonicity of $((q_1, \dots, q_{k+1}), (w_1, \dots, w_{k+1}))$ is guaranteed by optimality of the greedy solution for the single value $c = mw_k$. Moreover, this check does not require exact solution of the instance, since execution of the greedy algorithm for the value $c = \gamma$ (requiring $O(k)$ time) is sufficient to test condition (iii).

Given q_j, w_j ($j = 1, \dots, n$) satisfying (5.14) and (5.15), the following procedure checks condition (iii) for $k = 1, \dots, n - 1$. Note that condition (5.15), always satisfied by CMP, is not necessarily satisfied by an instance of UEMK.

```
procedure MNT:
input: n, (q_j), (w_j);
output: opt;
begin
    ñ := n;
    n := 1;
    optimal := "yes";
    while n < ñ and optimal = "yes" do
        begin
            m := ⌈w_{n+1}/w_n⌉;
```

```
            c := mw_n − w_{n+1};
            call GREEDYUM;
            if q_{n+1} + z^g ≤ mq_n then n := n + 1 else optimal := "no"
        end;
    opt := n;
    n := ñ
end.
```

The time complexity of MNT is clearly $O(n^2)$. On return, we know that the pairs $((q_1, \ldots, q_k), (w_1, \ldots, w_k))$ are canonical for all $k \leq opt$. If $opt < n$, then the pair with $k = opt + 1$ is not canonical, while, for the pairs with $k > opt + 1$ the situation is undecided.

Example 5.2 (continued)

By setting $q_j = 1$ for $j = 1, \ldots, 6$, and applying MNT, we have

$n = 1, 2, 3 : m = 2, c = 0, z^g = 0;$

$n = 4 \qquad : m = 2, c = 6, z^g = 2, opt = 4.$

Hence the greedy algorithm exactly solves all the instances induced by items $(1, \ldots, k)$ with $k \leq 4$, while it fails for at least one instance with $k = 5$ (see, e.g., the case with $c = 16$). The situation for $(w_1, \ldots, w_6)$ cannot be decided through procedure MNT, although the vector is canonical, as can be proved using Theorem 5.2 . □

Further characterizations of instances for which the greedy solution is optimal have been given by Chang and Korsh (1976) and Tien and Hu (1977).

5.5 EXACT ALGORITHMS

Chang and Gill (1970a) have presented a recursive procedure for the exact solution of those instances of CMP in which one item type has weight of value 1. An Algol implementation of this method has been given by Chang and Gill (1970b) and corrected by Johnson and Kernighan (1972). The resulting code is, however, highly inefficient, as shown in Martello and Toth (1977c, 1980b) through computational experiments, so no account of it will be taken in the following.

In the following sections we consider algorithms for the exact solution of CMP with no special assumption.

5.5.1 Dynamic programming

Given a pair of integers m $(1 \leq m \leq n)$ and $\hat{c}$ $(0 \leq \hat{c} \leq c)$, consider the sub-instance of CMP consisting of item types $1, \ldots, m$ and capacity $\hat{c}$, and denote with $f_m(\hat{c})$ the corresponding optimal solution value ($f_m(\hat{c}) = \infty$ if no solution of value $\hat{c}$ exists). Then, clearly,

$$f_1(\hat{c}) = \begin{cases} \infty & \text{for all positive } \hat{c} \le c \text{ such that } \hat{c}(\bmod w_1) \neq 0; \\ l & \text{for } \hat{c} = lw_1, \text{ with } l = 0, \dots, \left\lfloor \dfrac{c}{w_1} \right\rfloor. \end{cases}$$

$f_m(\hat{c})$ can be computed, by considering increasing values of m from 2 to n and, for each m, increasing values of $\hat{c}$ from 0 to c, as

$$f_m(\hat{c}) = \min \left\{ f_{m-1}(\hat{c} - lw_m) + l : l \text{ integer}, \ 0 \le l \le \left\lfloor \frac{\hat{c}}{w_m} \right\rfloor \right\}.$$

The optimal solution value of CMP is then given by $f_n(c)$. The time complexity for this computation is $O(nc^2)$, the space complexity $O(nc)$.

By adapting to CMP the improved recursion proposed by Gilmore and Gomory (1965) for the unbounded knapsack problem (Section 3.6.2), we obtain

$$f_m(\hat{c}) = \begin{cases} f_{m-1}(\hat{c}) & \text{for } \hat{c} = 0, \dots, w_m - 1; \\ \min\left(f_{m-1}(\hat{c}), f_m(\hat{c} - w_m) + 1 \right) & \text{for } \hat{c} = w_m, \dots, c, \end{cases}$$

which reduces the time complexity to $O(nc)$. Wright (1975) has further noted that, if the items are sorted according to increasing weights, only values of $\hat{c}$ not greater than $w_m w_{m+1}$ need be considered at each stage m. In fact, w_m items of type $m+1$ give the same weight as w_{m+1} items of type m and a better value for the objective function. A specialized dynamic programming algorithm for CMP can be found in Wright (1975).

5.5.2 Branch-and-bound

In the present section we assume that the item types are sorted so that

$$w_1 > w_2 > \dots > w_n. \tag{5.19}$$

To our knowledge, the only branch-and-bound algorithms for CMP are those in Martello and Toth (1977c, 1980b). We give a description of the latter, which has a structure similar to that of MTU1 (Section 3.6.2), with one important difference.

As in MTU1, at a forward move associated with the jth item type, if a lower bound on the best solution obtainable is less than the best solution so far, the largest possible number of items of type j is selected. As usual, a backtracking move consists in removing one item of the last inserted type. Before backtracking on item type i, let $\hat{x}_j$ ($j = 1, \dots, i$) be the current solution, $\hat{c} = c - \sum_{j=1}^{i} w_j \hat{x}_j$ and $\hat{z} = \sum_{j=1}^{i} \hat{x}_j$ the corresponding residual capacity and value, and z the best solution value so far. The following is then true (Martello and Toth, 1980b):

if $\hat{c} < w_i$, *the value* $l_{\hat{c}} = z - \hat{z}$ *is a lower bound on* $f_n(\hat{c})$ (= number of items needed to obtain a change $\hat{c}$ with item weights $(w_1, \ldots, w_n)$, see Section 5.5.1). In fact: (a) only item types $i+1, \ldots, n$ can produce $\hat{c}(< w_i)$, so (b) if the solution of value z has been obtained at a decision node descending from the current one, then, clearly, $l_{\hat{c}} = f_n(\hat{c})$; otherwise, at each leaf λ of the decision sub-tree descending from the current node, the lower bound, say $\hat{z} + L_\lambda$, allowed the search to be stopped, so a valid lower bound on $f_n(\hat{c})$ is $\min_\lambda \{L_\lambda\} \geq z - \hat{z} = l_{\hat{c}}$.

The consideration above leads to a dominance criterion which is particularly strong, since it allows one to fathom nodes of the decision tree basing oneself on a value depending only on the current residual capacity, independently of the variables currently fixed. In the resulting algorithm, $l_{\hat{c}}$ is defined at Step 5, and tested at Steps 2a and 5.

Also, it is useful to initially define a vector $(m_{\hat{c}})$ such that $m_{\hat{c}} = \min \{j : w_j \leq \hat{c}\}$, so that, for each residual capacity $\hat{c}$ produced by branch-and-bound, the next feasible item type is immediately known. Vectors $(l_{\hat{c}})$ and $(m_{\hat{c}})$ clearly require pseudo-polynomial space, hence, in the following implementation, their size is limited by a constant parameter γ. It is assumed that the item types are sorted according to (5.19), and that $\gamma < w_1$. (Note that vector $(m_{\hat{c}})$ can be used only after a forward move, i.e. when $\hat{c} < w_1$, while after a backtracking, say on item type i, the next feasible item type is $i+1$.)

procedure MTC1:
input: $n, c, (w_j), \gamma$;
output: $z, (x_j)$;
begin
1. [initialize]
   $z := c + 1$;
   $w_{n+1} := 1$;
   for $k := 1$ **to** n **do** $\hat{x}_k := 0$;
   compute the lower bound $L = L_2$ (Section 5.2) on the optimal solution value;
   $j := n$;
   while $j > 1$ and $w_j \leq \gamma$ **do**
      begin
         for $h := w_j$ **to** $\min(\gamma, w_{j-1} - 1)$ **do** $m_h := j$;
         $j := j - 1$
      end;
   for $h := 1$ **to** $\min(\gamma, w_n - 1)$ **do** $l_h := \infty$;
   for $h := w_n$ **to** γ **do** $l_h := 0$;
   $\hat{x}_1 := \lfloor c/w_1 \rfloor$;
   $\hat{z} := \hat{x}_1$;
   $\hat{c} := c - w_1\hat{x}_1$;
   $j := 2$;
   if $\hat{c} > 0$ **then go to** 2a;
   $z := \hat{x}_1$;
   for $j := 1$ **to** n **do** $x_j := \hat{x}_j$;
   return ;
2a. [dominance criterion]

```
        if ĉ ≤ γ then
              if l_ĉ ≥ z − ẑ then go to 5 else j := m_ĉ
        else
              if ĉ < w_n then go to 5 else find j = min{k ≥ j : w_k ≤ ĉ};
    2. [build a new current solution]
        y := ⌊ĉ/w_j⌋;
        c̃ := ĉ − y w_j;
        if z ≤ ẑ + y + ⌈c̃/w_{j+1}⌉ then go to 5;
        if c̃ = 0 then go to 4;
        if j = n then go to 5;
    3. [save the current solution]
        ĉ := c̃;
        ẑ := ẑ + y;
        x̂_j := y;
        j := j + 1;
        go to 2a;
    4. [update the best solution so far]
        z := ẑ + y;
        for k := 1 to j − 1 do x_k := x̂_k;
        x_j := y;
        for k := j + 1 to n do x_k := 0;
        if z = L then return;
    5. [backtrack]
        find i = max{k < j : x̂_k > 0};
        if no such i then return ;
        if ĉ ≤ min(γ, w_i − 1) then l_ĉ := max(l_ĉ, z − ẑ);
        ĉ := ĉ + w_i;
        ẑ := ẑ − 1;
        x̂_i := x̂_i − 1;
        if z ≤ ẑ + ⌈ĉ/w_{i+1}⌉ then (comment: remove all items of type i)
            begin
                ĉ := ĉ + w_i x̂_i;
                ẑ := ẑ − x̂_i;
                x̂_i := 0;
                j := i;
                go to 5
            end;
        j := i + 1;
        if ĉ < γ and l_ĉ ≥ z − ẑ then go to 5;
        if ĉ − w_i ≥ w_n then go to 2;
        h := i;
    6. [try to replace one item of type i with items of type h]
        h := h + 1;
        if z ≤ ẑ + ⌈ĉ/w_h⌉ or h > n then go to 5;
        if ĉ − w_h < w_n then go to 6;
        j := h;
        go to 2
    end.
```

Example 5.1 (continued)

Recall the instance

$$n = 5,$$
$$(w_j) = (11, 8, 5, 4, 1),$$
$$c = 29.$$

Figure 5.1 gives the decision-tree produced by MTC1 (with $\gamma = 10$). □

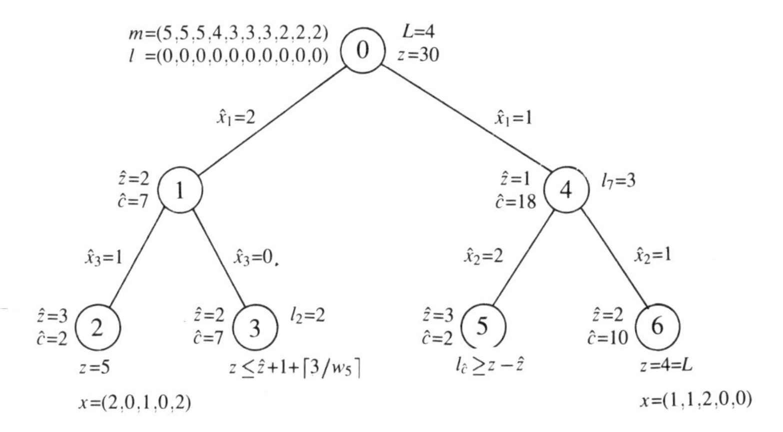

Figure 5.1 Decision-tree of procedure MTC1 for Example 5.1

The Fortran implementation of procedure MTC1 is included in that of procedure MTC2, which is described in the next section.

5.6 AN EXACT ALGORITHM FOR LARGE-SIZE PROBLEMS

Computational experiments with algorithm MTC1 (Martello and Toth, 1980b) show that, similarly to what happens for other knapsack-type problems (see Sections 2.9, 3.6.3, 4.2.3), many instances of CMP can be solved efficiently even for high values of n and, in such cases, the number of item types used for the solution is considerably smaller than n.

For CMP, however, the *core problem* does not consist of the most "favourable" item types (those with highest weight), since the equality constraint often forces the optimal solution to include also some items with medium and small weight.

An experimental analysis of the structure of the solutions found by MTC1 shows two facts: (a) the optimal solution value is often equal to the value of bound L_2 (Section 5.2); (b) many equivalent solutions usually exist. Hence we define the *core* as

$$C = \{j_1, \dots, j_{\overline{n}}\},$$

with

j_1, j_2, j_3 = the three item types of maximum weight,

$j_4, \dots, j_{\overline{n}}$ = any $\overline{n} - 3$ other item types,

and the *core problem* as

$$\text{minimize} \quad z = \sum_{j \in C} x_j$$

$$\text{subject to} \quad \sum_{j \in C} w_j x_j = c,$$

$$x_j \geq 0 \text{ and integer}, \qquad j \in C.$$

Noting that j_1, j_2 and j_3 are the only item types needed to compute L_2, the following procedure attempts to solve CMP by sorting only a subset of the item types. In input, $\overline{n} \leq n - 3$ is the expected size of the core, γ the parameter needed by MTC1.

procedure MTC2:
input: $n, c, (w_j), \gamma, \overline{n}$;
output: $z, (x_j)$;
begin
  determine the three item types (j_1, j_2, j_3) of maximum weight;
  compute lower bound L_2;
  $C := \{1, \dots, \overline{n}\} \cup \{j_1, j_2, j_3\}$;
  sort the item types in C in order of decreasing weights;
  solve the core problem exactly, using MTC1, and let z and (x_j) define the solution;
  if $z = L_2$ **then for each** $j \in \{1, \dots, n\} \backslash C$ **do** $x_j := 0$
  else
    begin
      sort item types $1, \dots, n$ in order of decreasing weights;
      solve CMP exactly, using MTC1, and let z and (x_j) define the solution
    end
end.

"Good" values for $\overline{n}$ and γ were experimentally determined as

$$\overline{n} = \min\left(n, \max\left(500, \left\lfloor \frac{n}{20} \right\rfloor\right)\right),$$

$$\gamma = \min(10\,000, w_1 - 1).$$

The Fortran implementation of algorithm MTC2 is included in the present volume.

5.7 COMPUTATIONAL EXPERIMENTS

In the present section we analyse the computational behaviour of the Fortran IV implementations of algorithms for CMP on data sets having

$$w_j \text{ uniformly random in } [1, M].$$

In Table 5.1 we compare the dynamic programming approach of Wright (Section 5.5.1), algorithm MTC1 (Section 5.5.2) and the approximate algorithm GREEDYUM (Section 5.3) on problems with $M = 4n$, for two values of c. The recursive approach of Chang and Gill (Section 5.5) is not considered since computational experiments (Martello and Toth, 1980b) showed that it is very much slower than the Wright algorithm. For each value of n and c, we generated 100 problems admitting of a feasible solution. Each algorithm had a time limit of 250 seconds to solve them. The entries give the average running times (including sorting times) obtained by the three algorithms on a CDC Cyber 730, the percentages of approximate solutions which are sub-optimal, infeasible and optimal, respectively, and the average running time of MTC1 when GREEDYUM finds the optimum. The table shows that MTC1 clearly dominates the Wright algorithm. The greedy algorithm is about twice as fast as MTC1, but the quality of its solutions is rather poor. In addition, for the instances in which it gives the optimal solution, the running

Table 5.1 w_j uniformly random in $[1, 4n]$. CDC-Cyber 730 in seconds. Average values over 100 feasible problems

					Greedy solutions			
c	n	Wright (time)	MTC1 (time)	Greedy (time)	Sub-optimal (%)	Not feasible (%)	Optimal (%)	MTC1 When Greedy is optimal (time)
$10n$	25	0.135	0.006	0.002	35	43	22	0.004
	50	0.451	0.011	0.006	35	38	27	0.009
	100	1.612	0.022	0.013	35	47	18	0.018
	200	time	0.045	0.029	40	42	18	0.035
	500	—	0.119	0.078	42	35	23	0.098
	1 000	—	0.241	0.155	33	40	27	0.202
$\sum_{j=1}^{n} w_j$	25	0.166	0.006	0.002	35	43	22	0.004
	50	0.518	0.011	0.006	32	45	23	0.009
	100	1.768	0.022	0.013	29	44	27	0.018
	200	time	0.045	0.030	40	37	27	0.035
	500	—	0.121	0.078	29	43	28	0.096
	1 000	—	0.240	0.154	24	41	35	0.204

time of MTC1 is only slightly higher. The running times of all the algorithms are practically independent of the value of c.

Table 5.2 gives the average times obtained, for larger problems, by MTC1 and MTC2 (Section 5.6) on an HP 9000/840 (with option "-o" for the Fortran compiler). The problems, not necessarily feasible, were generated with $M = 4n$ and $c = 0.5 \sum_{j=1}^{n} w_j$. The sorting times, included in the entries, are also separately given. The table shows that, for $n \geq 1\,000$, i.e. when the core problem is introduced, MTC2 is considerably faster than MTC1.

Table 5.3 analyses the behaviour of MTC2 when M varies, and shows that higher values of M tend to produce harder problems. This can be explained by noting that increasing the data scattering makes it more difficult to satisfy the equality constraint. Hence, in order to evaluate MTC1 and MTC2 on difficult problems, we set $M = 10^5$ for all values of n. Table 5.4 confirms the superiority of MTC2 also

Table 5.2 w_j uniformly random in [1, $4n$]; $c = 0.5 \sum_{j=1}^{n} w_j$. HP 9000/840 in seconds. Average times over 20 problems

n	Sorting	MTC1	MTC2
50	0.003	0.007	0.003
100	0.006	0.010	0.008
200	0.013	0.021	0.019
500	0.036	0.054	0.051
1 000	0.079	0.114	0.064
2 000	0.165	0.237	0.087
5 000	0.468	0.595	0.121
10 000	0.963	1.198	0.179
20 000	2.073	12.860	0.370

Table 5.3 Algorithm MTC2. w_j uniformly random in [1, M]; $c = 0.5 \sum_{j=1}^{n} w_j$. HP 9000/840 in seconds. Average times over 20 problems

n	$M = 4n$	$M = 8n$	$M = 12n$
100	0.008	0.013	0.017
1 000	0.064	0.081	0.096
10 000	0.179	0.309	4.743

Table 5.4 w_j uniformly random in [1, 10^5]; $c = 0.5 \sum_{j=1}^{n} w_j$. HP 9000/840 in seconds. Average times over 20 problems

n	Sorting	MTC1	MTC2
50	0.004	1.205	1.172
100	0.007	0.754	0.744
200	0.012	0.862	0.855
500	0.037	2.321	2.306
1 000	0.082	3.078	1.098
2 000	0.171	7.778	1.654
5 000	0.456	11.141	0.810
10 000	0.948	time	1.939
20 000	2.124	—	5.480

for this generation. Note that, for $n \geq 1\,000$, the time difference is considerably higher than the sorting time, indicating that MTC2 takes advantage of the lesser number of item types also in the branch-and-bound phase. For $n = 10\,000$, MTC1 could not solve the generated problems within 250 seconds.

5.8 THE BOUNDED CHANGE-MAKING PROBLEM

The *Bounded Change-Making Problem* (BCMP) is the generalization of CMP we obtain by introducing

$$b_j = \text{upper bound on the availability of items of type } j,$$

and formulating the problem as

$$\text{minimize} \quad z = \sum_{j=1}^{n} x_j \tag{5.20}$$

$$\text{subject to} \quad \sum_{j=1}^{n} w_j x_j = c, \tag{5.21}$$

$$0 \leq x_j \leq b_j \text{ and integer}, \qquad j = 1, \ldots, n. \tag{5.22}$$

We will maintain the assumptions made in Section 5.1. In addition we will assume, without loss of generality, that values b_j (positive integer for all j) satisfy

$$\sum_{j=1}^{n} b_j w_j > c, \tag{5.23}$$

$$b_j w_j \leq c, \qquad j = 1, \ldots, n. \tag{5.24}$$

Violation of assumption (5.23) produces an infeasible or trivial problem, while for each j not satisfying (5.24), we can replace b_j with $\lfloor c/w_j \rfloor$.

By assuming that the item types are sorted so that

$$w_1 > w_2 > \ldots > w_n, \tag{5.25}$$

the continuous relaxation of BCMP can easily be solved, as for the bounded knapsack problem, through a straightforward adaptation of Theorem 2.1. Define the *critical item type* s as

$$s = \min\left\{ j : \sum_{i=1}^{j} b_i w_i > c \right\},$$

and

$$\overline{c} = c - \sum_{j=1}^{s-1} b_j w_j. \tag{5.26}$$

Then the optimal solution $\overline{x}$ of the continuous relaxation is

$$\overline{x}_j = b_j \qquad \text{for } j = 1, \ldots, s-1,$$

$$\overline{x}_j = 0 \qquad \text{for } j = s+1, \ldots, n,$$

$$\overline{x}_s = \frac{\overline{c}}{w_s},$$

and the corresponding value produces a lower bound for BCMP:

$$LB_1 = \sum_{j=1}^{s-1} b_j + \left\lceil \frac{\overline{c}}{w_s} \right\rceil.$$

A tighter bound, conceptually close to bound L_2 of Section 5.2, can be obtained by noting that, in the optimal solution, either $x_s \leq \lfloor \overline{c}/w_s \rfloor$ or $x_s > \lfloor \overline{c}/w_s \rfloor$. By defining

$$z' = \sum_{j=1}^{s-1} b_j + \left\lfloor \frac{\overline{c}}{w_s} \right\rfloor, \tag{5.27}$$

$$c' = \overline{c} \pmod{w_s}, \tag{5.28}$$

the respective lower bounds are

$$LB^0 = z' + \left\lceil \frac{c'}{w_{s+1}} \right\rceil, \tag{5.29}$$

$$LB^1 = z' - 1 + \left\lceil \frac{c' + w_{s-1}}{w_s} \right\rceil. \tag{5.30}$$

Hence,

Theorem 5.4 (Martello and Toth, 1977c) *The value*

$$LB_2 = \min\,(LB^0, LB^1),$$

where LB^0 and LB^1 are defined by (5.26)–(5.30), *is a lower bound for BCMP.*

LB_2 clearly dominates LB_1, since $LB_1 = z' + \lceil c'/w_s \rceil \leq LB^0$ and $LB^1 = z' + \lceil (c' + w_{s-1} - w_s)/w_s \rceil \geq LB_1$. The time complexity for the computation of LB_1

or LB_2 is $O(n)$ if the item types are already sorted according to (5.25). If this is not the case, the computation can still be done in $O(n)$ time, through an adaptation of procedure CRITICAL_ ITEM of Section 2.2.2.

A greedy algorithm for BCMP immediately derives from procedure GREEDYUM of Section 5.3, by replacing the definition of x_j and z^g with $x_j := \min(\lfloor \overline{c}/w_j \rfloor, b_j)$ and $z^g := z^g + x_j$, respectively. In this case too, the worst-case behaviour is arbitrarily bad, as shown by the counterexample of Section 5.3 with $b_1 = k$, $b_2 = 2$, $b_3 = 1$. To our knowledge, no further theoretical result on the behaviour of greedy algorithms for BCMP is known.

Exact algorithms for BCMP can also be obtained easily from those for CMP. In particular, a branch-and-bound algorithm MTCB derives from procedure MTC1 of Section 5.5.2 as follows. Apart from straightforward modifications at Step 1 ($b_{n+1} := +\infty$; $L = LB_2$; $\hat{x}_1 := \min(\lfloor c/w_1 \rfloor, b_1)$; **if** $\hat{c} > 0$ **then go to** 2), the main modification concerns Steps 2 and 4. For BCMP it is worth building a new current solution by inserting as many items of types $j, j+1, \ldots$ as allowed by their bounds, until the first is found whose bound cannot be reached. In order to avoid useless backtrackings, this solution is saved only if its value does not represent the minimum which can be obtained with item type j. The altered steps are then as follows.

```
2. [build a new current solution]
     y' := 0;
     c̃ := ĉ;
     i := j − 1;
     repeat
          i := i + 1;
          y := min(⌊c̃/w_i⌋, b_i);
          y' := y' + y;
          c̃ := c̃ − y w_i;
          if y = b_i then w̄ := w_{i+1} else w̄ := w_i;
          if z ≤ ẑ + y' + ⌈c̃/w̄⌉ then go to 5;
          if c̃ = 0 then go to 4;
          if i = n then go to 5
     until y < b_i;
     ẑ := ẑ + (y' − y);
     for k := j to i − 1 do x̂_k := b_k;
     j := i;

4. [update the best solution so far]
     z := ẑ + y';
     for k := 1 to j − 1 do x_k := x̂_k;
     for k := j to i − 1 do x_k := b_j;
     x_i := y;
     for k := i + 1 to n do x_k := 0;
     if z = L then return;
```

The Fortran code of algorithm MTCB is included in the present volume.

Table 5.5 Algorithm MTCB. $c = 0.5\sum_{j=1}^{n} w_j$. HP 9000/840 in seconds. Average times over 20 problems

	w_j uniformly random in [1, 4n]		w_j uniformly random in [1, 10^5]	
n	b_j in [1, 5]	b_j in [1, 10]	b_j in [1, 5]	b_j in [1, 10]
50	0.009	0.010	1.646	1.442
100	0.016	0.019	1.230	1.033
200	0.038	0.036	1.073	0.934
500	0.100	0.099	1.233	2.051
1000	0.213	0.210	9.894	11.377
2000	0.453	0.449	6.145	20.145
5000	1.207	1.201	18.622	35.799
10000	2.429	2.377	—	—

Table 5.5 gives the results of computational experiments performed on the data generation of Tables 5.2 and 5.4, with values b_j uniformly random in ranges [1, 5] and [1, 10]. The solution of BCMP appears harder than that of CMP. The times in the first two columns are approximately twice as high as those obtained by MTC1 in Table 5.2. Increasing the range of values b_j did not alter the difficulty. The times in the third column are higher than those of MTC1 in Table 5.4. Larger values of b_j considerably increased in this case the difficulty, especially for large values of n.

6

0-1 Multiple knapsack problem

6.1 INTRODUCTION

The *0-1 Multiple Knapsack Problem* (MKP) is: given a set of n *items* and a set of m *knapsacks* ($m \leq n$), with

$$p_j = \textit{profit} \text{ of item } j,$$

$$w_j = \textit{weight} \text{ of item } j,$$

$$c_i = \textit{capacity} \text{ of knapsack } i,$$

select m disjoint subsets of items so that the total profit of the selected items is a maximum, and each subset can be assigned to a different knapsack whose capacity is no less than the total weight of items in the subset. Formally,

$$\text{maximize} \quad z = \sum_{i=1}^{m} \sum_{j=1}^{n} p_j x_{ij} \tag{6.1}$$

$$\text{subject to} \quad \sum_{j=1}^{n} w_j x_{ij} \leq c_i, \qquad i \in M = \{1, \ldots, m\}, \tag{6.2}$$

$$\sum_{i=1}^{m} x_{ij} \leq 1, \qquad j \in N = \{1, \ldots, n\}, \tag{6.3}$$

$$x_{ij} = 0 \text{ or } 1, \qquad i \in M, j \in N, \tag{6.4}$$

where

$$x_{ij} = \begin{cases} 1 & \text{if item } j \text{ is assigned to knapsack } i; \\ 0 & \text{otherwise.} \end{cases}$$

When $m = 1$, MKP reduces to the 0-1 (single) knapsack problem considered in Chapter 2.

We will suppose, as is usual, that the weights w_j are positive integers. Hence, without loss of generality, we will also assume that

$$p_j \text{ and } c_i \text{ are positive integers,} \tag{6.5}$$

$$w_j \le \max_{i \in M}\{c_i\} \quad \text{for } j \in N, \tag{6.6}$$

$$c_i \ge \min_{j \in N}\{w_j\} \quad \text{for } i \in M, \tag{6.7}$$

$$\sum_{j=1}^{n} w_j > c_i \quad \text{for } i \in M. \tag{6.8}$$

If assumption (6.5) is violated, fractions can be handled by multiplying through by a proper factor, while nonpositive values can easily be handled by eliminating all items with $p_j \le 0$ and all knapsacks with $c_i \le 0$. (There is no easy way, instead, of transforming an instance so as to handle negative weights, since the Glover (1965) technique given in Section 2.1 does not extend to MKP. All the considerations in this Chapter, however, easily extend to the case of nonpositive values.) Items j violating assumption (6.6), as well as knapsacks i violating assumption (6.7), can be eliminated. If a knapsack, say i^*, violates assumption (6.8), then the problem has the trivial solution $x_{i^*j} = 1$ for $j \in N$, $x_{ij} = 0$ for $i \in M\backslash\{i^*\}$ and $j \in N$. Finally, observe that if $m > n$ then the $(m - n)$ knapsacks of smallest capacity can be eliminated.

We will further assume that the items are sorted so that

$$\frac{p_1}{w_1} \ge \frac{p_2}{w_2} \ge \ldots \ge \frac{p_n}{w_n}. \tag{6.9}$$

In Section 6.2 we examine the relaxation techniques used for determining upper bounds. Approximate algorithms are considered in Sections 6.3 and 6.6 . In Section 6.4 we describe branch-and-bound algorithms, in Section 6.5 reduction techniques. The results of computational experiments are presented in Section 6.7.

6.2 RELAXATIONS AND UPPER BOUNDS

Two techniques are generally employed to obtain upper bounds for MKP: the surrogate relaxation and the Lagrangian relaxation. As we show in the next section, the continuous relaxation of the former also gives the value of the continuous relaxation of MKP.

6.2.1 Surrogate relaxation

Given a positive vector $(\pi_1, \ldots, \pi_m)$ of multipliers, the standard *surrogate relaxation*, $S(MKP, \pi)$, of MKP is

$$\text{maximize} \quad \sum_{i=1}^{m}\sum_{j=1}^{n} p_j x_{ij} \tag{6.10}$$

$$\text{subject to} \quad \sum_{i=1}^{m} \pi_i \sum_{j=1}^{n} w_j x_{ij} \leq \sum_{i=1}^{m} \pi_i c_i \tag{6.11}$$

$$\sum_{i=1}^{m} x_{ij} \leq 1, \qquad j \in N, \tag{6.12}$$

$$x_{ij} = 0 \text{ or } 1, \qquad i \in M, j \in N. \tag{6.13}$$

Note that we do not allow any multiplier, say $\pi_{\bar{\imath}}$, to take the value zero, since this would immediately produce a useless solution ($x_{\bar{\imath}j} = 1$ for $j \in N$) of value $\sum_{j=1}^{n} p_j$. The optimal vector of multipliers, i.e. the one producing the minimum value of $z(S(MKP, \pi))$ and hence the tightest upper bound for MKP, is then defined by the following

Theorem 6.1 (Martello and Toth, 1981a) *For any instance of MKP, the optimal vector of multipliers for $S(MKP, \pi)$ is $\pi_i = k$ (k any positive constant) for all $i \in M$.*

Proof. Let $\bar{\imath} = \arg\min\{\pi_i : i \in M\}$, and suppose that (x_{ij}^*) defines an optimal solution to $S(MKP, \pi)$. A feasible solution of the same value can be obtained by setting $x_{ij}^* = 0$ and $x_{\bar{\imath}j}^* = 1$ for each $j \in N$ such that $x_{ij}^* = 1$ and $i \neq \bar{\imath}$ (since the only effect is to decrease the left-hand side of (6.11)). Hence $S(MKP, \pi)$ is equivalent to the 0-1 single knapsack problem

$$\text{maximize} \quad \sum_{j=1}^{n} p_j x_{\bar{\imath}j}$$

$$\text{subject to} \quad \sum_{j=1}^{n} w_j x_{\bar{\imath}j} \leq \left\lfloor \sum_{i=1}^{m} \pi_i c_i / \pi_{\bar{\imath}} \right\rfloor,$$

$$x_{\bar{\imath}j} = 0 \text{ or } 1, \qquad j \in N.$$

Since $\lfloor \sum_{i=1}^{m} \pi_i c_i / \pi_{\bar{\imath}} \rfloor \geq \sum_{i=1}^{m} c_i$, the choice $\pi_i = k$ (k any positive constant) for all $i \in M$ produces the minimum capacity and hence the minimum value of $z(S(MKP, \pi))$. □

By setting $\pi_i = k > 0$ for all $i \in M$, and $y_j = \sum_{i=1}^{m} x_{ij}$ for all $j \in N$, $S(MKP, \pi)$ becomes

$$\text{maximize} \quad \sum_{j=1}^{n} p_j y_j$$

$$\text{subject to} \quad \sum_{j=1}^{n} w_j y_j \leq \sum_{i=1}^{m} c_i,$$

$$y_j = 0 \text{ or } 1, \qquad j \in N,$$

which we denote simply with $S(MKP)$ in the following. Loosely speaking, this relaxation consists in using only one knapsack, of capacity

$$c = \sum_{i=1}^{m} c_i. \tag{6.14}$$

The computation of upper bound $z(S(MKP))$ for MKP has a non-polynomial time complexity, although many instances of the 0-1 knapsack problem can be solved very quickly, as we have seen in Chapter 2. Weaker upper bounds, requiring $O(n)$ time, can however be computed by determining any upper bound on $z(S(MKP))$ through the techniques of Sections 2.2 and 2.3.

A different upper bound for MKP could be computed through its continuous relaxation, $C(MKP)$, given by (6.1), (6.2), (6.3) and

$$0 \leq x_{ij} \leq 1, \qquad i \in M, \ j \in N. \tag{6.15}$$

This relaxation, however, is dominated by any of the previous ones, since it can be proved that its value is equal to that of the continuous relaxation of the surrogate relaxation of the problem, i.e.

Theorem 6.2 $z(C(MKP)) = z(C(S(MKP)))$.

Proof. It is clear that, by setting $\pi_i = k > 0$ for all i, $C(S(MKP))$, which is obtained from (6.10)–(6.13) by relaxing (6.13) to (6.15), coincides with $S(C(MKP))$, which is obtained from (6.1), (6.2), (6.3) and (6.15) by relaxing (6.2) to (6.11). Hence we have $z(C(S(MKP))) \geq z(C(MKP))$. We now prove that $z(C(MKP)) \geq z(C(S(MKP)))$ also holds.

The exact solution $(\bar{y}_j)$ of the continuous relaxation of $S(MKP)$ can easily be determined as follows. If $\sum_{j=1}^{n} w_j \leq c$, where c is given by (6.14), then $\bar{y}_j = 1$ for $j = 1, \ldots, n$ and $z(C(S(MKP))) = \sum_{j=1}^{n} p_j$. Otherwise, from Theorem 2.1,

$$\bar{y}_j = 1 \quad \text{for } j = 1, \ldots, s-1,$$

$$\bar{y}_j = 0 \quad \text{for } j = s+1, \ldots, n,$$

$$\bar{y}_s = \left(c - \sum_{j=1}^{s-1} w_j\right) / w_s,$$

where

$$s = \min\left\{j : \sum_{k=1}^{j} w_k > c\right\}, \tag{6.16}$$

and

$$z(C(S(MKP))) = \sum_{j=1}^{s-1} p_j + \left(c - \sum_{j=1}^{s-1} w_j\right) p_s / w_s. \tag{6.17}$$

It is now easy to show that there exists a feasible solution $(\bar{x}_{ij})$ to $C(MKP)$ for which $\sum_{i=1}^{m} \bar{x}_{ij} = \bar{y}_j$ for all $j \in N$. Such a solution, in fact, can be determined by consecutively inserting items $j = 1, 2, \ldots$ into knapsack 1 (and setting $\bar{x}_{1,j} = 1$, $\bar{x}_{ij} = 0$ for $i \neq 1$), until the first item, say j^*, is found which does not fit since the residual capacity $\bar{c}_1$ is less than w_{j^*}. We then insert the maximum possible fraction of w_{j^*} into knapsack 1 (by setting $\bar{x}_{1,j^*} = \bar{c}_1 / w_{j^*}$) and continue with the residual weight $\bar{w}_{j^*} = w_{j^*} - \bar{c}_1$ and knapsack 2, and so on. Hence $z(C(MKP)) \geq z(C(S(MKP)))$. □

Example 6.1

Consider the instance of MKP defined by

$n = 6;$

$m = 2;$

$(p_j) = (110, 150, 70, 80, 30, 5);$

$(w_j) = (40, 60, 30, 40, 20, 5);$

$(c_i) = (65, 85).$

The surrogate relaxation is the 0-1 single knapsack problem defined by (p_j), (w_j) and $c = 150$. Its optimal solution can be computed through any of the exact algorithms of Chapter 2:

$(x_j) = (1, 1, 1, 0, 1, 0)$, $z(S(MKP)) = 360$.

Less tight values can be computed, in $O(n)$ time, through any of the upper bounds of Sections 2.2, 2.3. Using the Dantzig (1957) bound (Theorem 2.1), we get

$s = 4$, $(\overline{x}_j) = (1, 1, 1, \frac{1}{2}, 0, 0)$, $U_1 = 370$ $(= z(C(MKP)))$.

This is also the value produced by the continuous relaxation of the given problem since, following the proof of Theorem 6.2, we can obtain, from $(\overline{x}_j)$,

$(\overline{x}_{1,j}) = (1, \frac{5}{12}, 0, 0, 0, 0)$,

$(\overline{x}_{2,j}) = (0, \frac{7}{12}, 1, \frac{1}{2}, 0, 0)$.

Using the Martello and Toth (1977a) bound (Theorem 2.2), we get $U_2 = 363$. □

6.2.2 Lagrangian relaxation

Given a vector $(\lambda_1, \ldots, \lambda_n)$ of nonnegative multipliers, the *Lagrangian relaxation* $L(MKP, \lambda)$ of MKP is

$$\text{maximize} \quad \sum_{i=1}^{m} \sum_{j=1}^{n} p_j x_{ij} - \sum_{j=1}^{n} \lambda_j \left(\sum_{i=1}^{m} x_{ij} - 1 \right) \tag{6.18}$$

$$\text{subject to} \quad \sum_{j=1}^{n} w_j x_{ij} \le c_i, \qquad i \in M, \tag{6.19}$$

$$x_{ij} = 0 \text{ or } 1, \qquad i \in M, j \in N. \tag{6.20}$$

Since (6.18) can be written as

$$\text{maximize} \quad \sum_{i=1}^{m} \sum_{j=1}^{n} \tilde{p}_j x_{ij} + \sum_{j=1}^{n} \lambda_j, \tag{6.21}$$

where

$$\tilde{p}_j = p_j - \lambda_j, \qquad j \in N, \tag{6.22}$$

the relaxed problem can be decomposed into a series of m independent 0-1 single knapsack problems (KP_i, $i = 1, \ldots, m$), of the form

$$\text{maximize} \quad z_i = \sum_{j=1}^{n} \tilde{p}_j x_{ij}$$

$$\text{subject to} \quad \sum_{j=1}^{n} w_j x_{ij} \leq c_i,$$

$$x_{ij} = 0 \text{ or } 1, \qquad j \in N.$$

Note that all these problems have the same vectors of profits and weights, so the only difference between them is given by the capacity. Solving them, we obtain the solution of $L(MKP, \lambda)$, of value

$$z(L(MKP, \lambda)) = \sum_{i=1}^{m} z_i + \sum_{j=1}^{n} \lambda_j. \tag{6.23}$$

For the Lagrangian relaxation there is no counterpart of Theorem 6.1, i.e. it is not known how to determine analytically the vector (λ_j) producing the lowest possible value of $z(L(MKP, \lambda))$. An approximation of the optimum (λ_j) can be obtained through subgradient optimization techniques which are, however, generally time consuming. Hung and Fisk (1978) were the first to use this relaxation to determine upper bounds for MKP, although Ross and Soland (1975) had used a similar approach for the generalized assignment problem (see Section 7.2.1), of which MKP is a particular case. They chose for (λ_j) the optimal dual variables associated with constraints (6.3) in $C(MKP)$. Using the complementary slackness conditions, it is not difficult to check that such values are

$$\overline{\lambda}_j = \begin{cases} p_j - w_j \dfrac{p_s}{w_s} & \text{if } j < s; \\ 0 & \text{if } j \geq s, \end{cases} \tag{6.24}$$

where s is the critical item of $S(MKP)$, defined by (6.14) and (6.16). (For $S(MKP)$, Hung and Fisk (1978) used the same idea, previously suggested by Balas (1967) and Geoffrion (1969), choosing for (π_i) the optimal dual variables associated with constraints (6.2) in $C(MKP)$, i.e. $\overline{\pi}_i = p_s/w_s$ for all i. Note that, on the basis of Theorem 6.1, this is an optimal choice.)

With choice (6.24), in each KP_i $(i = 1, \dots, m)$ we have $\tilde{p}_j/w_j = p_s/w_s$ for $j \leq s$ and $\tilde{p}_j/w_j \leq p_s/w_s$ for $j > s$. It follows that $z(C(L(MKP, \overline{\lambda}))) = (p_s/w_s)\sum_{i=1}^{m} c_i + \sum_{j=1}^{n} \overline{\lambda}_j$, so from (6.17) and Theorem 6.2,

$$z(C(L(MKP, \overline{\lambda}))) = z(C(S(MKP))) = z(C(MKP)),$$

i.e. both the Lagrangian relaxation with multipliers $\overline{\lambda}_j$ and the surrogate relaxation with multipliers $\pi_i = k > 0$ for all i, dominate the continuous relaxation. No dominance exists, instead, between them.

Computing $z(L(MKP, \overline{\lambda}))$ requires a non-polynomial time, but upper bounds on this value, still dominating $z(C(MKP))$, can be determined in polynomial time, by using any upper bound of Sections 2.2 and 2.3 for the m 0-1 single knapsack problems generated.

Example 6.1 (continued)

From (6.24), we get

$$(\overline{\lambda}_j) = (30, 30, 10, 0, 0, 0), (\tilde{p}_j) = (80, 120, 60, 80, 30, 5).$$

By exactly solving KP_1 and KP_2, we have

$$(x_{1,j}) = (0, 1, 0, 0, 0, 1), z_1 = 125,$$

$$(x_{2,j}) = (1, 0, 0, 1, 0, 1), z_2 = 165.$$

Hence $z(L(MKP, \overline{\lambda})) = 360$, i.e. the Lagrangian and surrogate relaxation produce the same value in this case.

By using U_1 or U_2 (see Sections 2.2.1 and 2.3.1) instead of the optimal solution values, the upper bound would result in 370 (= 130 + 170 + 70). □

It is worth noting that feasibility of the solution of $L(MKP, \lambda)$ for MKP can easily be verified, in $O(nm)$ time, by checking conditions (6.3) (for the example above, $x_{1,6} + x_{2,6} \leq 1$ is not satisfied). This is not the case for $S(MKP)$, for which testing feasibility is an NP-complete problem. In fact, determining whether a subset of items can be inserted into knapsacks of given capacities generalizes the bin-packing problem (see Chapter 8) to the case in which containers of different capacity are allowed.

We finally note that a second Lagrangian relaxation is possible. For a given vector $(\mu_1, \ldots, \mu_m)$ of positive multipliers, $L(MKP, \mu)$ is

$$\text{maximize} \quad \sum_{i=1}^{m}\sum_{j=1}^{n} p_j x_{ij} - \sum_{i=1}^{m} \mu_i \left(\sum_{j=1}^{n} w_j x_{ij} - c_i \right) \tag{6.25}$$

$$\text{subject to} \quad \sum_{i=1}^{m} x_{ij} \leq 1, \qquad j \in N,$$

$$x_{ij} = 0 \text{ or } 1, \qquad i \in M, j \in N.$$

Note that, as in the case of $S(MKP, \pi)$, we do not allow any multiplier to take the value zero, which again would produce a useless solution value. By writing (6.25) as

$$\text{maximize} \sum_{i=1}^{m}\sum_{j=1}^{n}(p_j - \mu_i w_j)x_{ij} + \sum_{i=1}^{m}\mu_i c_i,$$

it is clear that the optimal solution can be obtained by determining $i^* = \arg\min\{\mu_i : i \in M\}$, and setting, for each $j \in N$: $x_{i^*j} = 1$ if $p_j - \mu_{i^*}w_j > 0$, $x_{i^*j} = 0$ otherwise, and $x_{ij} = 0$ for all $i \in M\backslash\{i^*\}$. Since this is also the optimal solution of $C(L(MKP, \mu))$, we have $z(L(MKP, \mu)) \geq z(C(MKP))$, i.e. this relaxation cannot produce, for MKP, a bound tighter than the continuous one. (Using $\overline{\mu}_i = p_s/w_s$ for all $i \in M$, we have $z(L(MKP, \overline{\mu})) = \sum_{j=1}^{s-1}(p_j - (p_s/w_s)w_j) + cp_s/w_s = z(C(MKP))$, with c and s given by (6.14) and (6.16), respectively.)

6.2.3 Worst-case performance of the upper bounds

We have seen that the most natural polynomially-computable upper bound for MKP is

$$U_1 = \lfloor z(C(MKP))\rfloor = \lfloor z(C(S(MKP)))\rfloor = \lfloor z(C(L(MKP, \overline{\lambda})))\rfloor.$$

Theorem 6.3 $\rho(U_1) = m + 1$.

Proof. We first prove that $\rho(U_1) \leq m + 1$, by showing that

$$z(C(S(MKP))) \leq (m+1)z(MKP).$$

Consider the solution of $C(S(MKP))$ and let us assume, by the moment, that $\sum_{j=1}^{n} w_j > \sum_{i=1}^{m} c_i$. Let s_i denote the *critical item relative to knapsack i* ($i \in M$) defined as

$$s_i = \min\left\{k : \sum_{j=1}^{k} w_j > \sum_{l=1}^{i} c_l\right\}. \tag{6.26}$$

Note that, from Theorem 6.2, the only fractional variable in the solution is $\overline{y}_s$, with $s \equiv s_m$. Hence the solution value can be written as

$$z(C(S(MKP))) = \sum_{j=1}^{s_1-1} p_j + p_{s_1} + \sum_{j=s_1+1}^{s_2-1} p_j + p_{s_2} + \ldots + \sum_{j=s_{m-1}+1}^{s_m-1} p_j$$

$$+ \left(c - \sum_{j=1}^{s-1} w_j\right)\frac{p_s}{w_s}; \tag{6.27}$$

from which we have the thesis, since

(a) Selecting items $\{s_{i-1}+1,\ldots,s_i-1\}$ (where $s_0 = 0$) for insertion into knapsack i $(i = 1,\ldots,m)$, we obtain a feasible solution for MKP, so $z(MKP) \geq \sum_{i=1}^{m} \sum_{j=s_{i-1}+1}^{s_i-1} p_j$;

(b) From assumption (6.6), $z(MKP) \geq p_{s_i}$ for all $i \in M$, hence also $z(MKP) \geq (c - \sum_{j=1}^{s-1} w_j)p_s/w_s$.

If $\sum_{j=1}^{n} w_j \leq \sum_{i=1}^{m} c_i$ the result holds a fortiori, since some terms of (6.27) are null.

To see that $m+1$ is tight, consider the series of instances with: $n \geq 2m$; $c_1 = 2k$ $(k \geq 2), c_2 = \ldots = c_m = k$; $p_1 = \ldots = p_{m+1} = k, w_1 = \ldots = w_{m+1} = k+1$; $p_{m+2} = \ldots = p_n = 1, w_{m+2} = \ldots = w_n = k$. We have $s \leq m+1, z(C(S(MKP))) = (m+1)k(k/(k+1)), z(MKP) = k+(m-1)$, so the ratio $U_1/z(MKP)$ can be arbitrarily close to $(m+1)$, for k sufficiently large. □

Any upper bound U, computable in polynomial time by applying the bounds of Sections 2.2 and 2.3 to $S(MKP)$ or to $L(MKP, \overline{\lambda})$, dominates U_1, hence $\rho(U) \leq m+1$. Indeed, this value is also tight, as can be verified through counter-examples obtained from that of Theorem 6.3 by adding a sufficiently large number of items with $p_j = k$ and $w_j = k+1$.

Finally note that, obviously, $\rho(U) \leq m+1$ also holds for those upper bounds U which can be obtained, in non-polynomial time, by exactly solving $S(MKP)$ or $L(MKP, \overline{\lambda})$.

6.3 GREEDY ALGORITHMS

As in the case of the 0-1 single knapsack problem (Section 2.4), also for MKP the continuous solution produces an immediate feasible solution, consisting (see (6.26), (6.27)) of the assignment of items $s_{i-1}+1,\ldots,s_i-1$ to knapsack i $(i = 1,\ldots,m)$ and having value

$$z' = \sum_{i=1}^{m} \sum_{j=s_{i-1}+1}^{s_i-1} p_j. \tag{6.28}$$

Since $z' \leq z \leq U_1 \leq z' + \sum_{i=1}^{m} p_{s_i}$, where $z = z(MKP)$, the absolute error of z' is less than $\sum_{i=1}^{m} p_{s_i}$. The worst-case relative error, instead, is arbitrarily bad, as shown by the series of instances with $n = 2m, c_i = k \geq m$ for $i = 1,\ldots,m$, $p_j = w_j = 1$ and $p_{j+1} = w_{j+1} = k$ for $j = 1,3,\ldots,n-1$, for which $z' = m$ and $z = mk$, so the ratio z'/z is arbitrarily close to 0 for k sufficiently large.

In this case too we can improve on the heuristic by also considering the solution consisting of the best critical item alone, i.e.

$$z^h = \max\left(z', \max_{i \in M}\{p_{s_i}\}\right).$$

The worst-case performance ratio of z^h is $1/(m+1)$. Since, in fact, $z^h \geq z'$ and $z^h \geq p_{s_i}$ for $i = 1, \ldots, m$, we have, from $z \leq z' + \sum_{i=1}^m p_{s_i}$, that $z \leq (m+1)z^h$. The series of instances with $n = 2m+2, c_1 = 2k (k > m), c_i = k$ for $i = 2, \ldots, m, p_j = w_j = 1$ and $p_{j+1} = w_{j+1} = k$ for $j = 1, 3, \ldots, n-1$ proves the tightness, since $z^h = k + m + 1$ and $z = (m+1)k$, so the ratio z^h/z is arbitrarily close to $1/(m+1)$ for k sufficiently large. Notice that the "improved" heuristic solution $z^g = \max(z', \max_{j \in N}\{p_j\})$ has the same worst-case performance.

For the heuristic solutions considered so far, value z' can be obtained, without solving $C(MKP)$, by an $O(n)$ greedy algorithm which starts by determining the critical item $s \equiv s_m$ through the procedure of Section 2.2.2, and re-indexing the items so that $j < s$ (resp. $j > s$) if $p_j/w_j > p_s/w_s$ (resp. $p_j/w_j < p_s/w_s$). Indices i and j are then initialized to 1 and the following steps are iteratively executed: (1) if $w_j \leq \overline{c}_i$ ($\overline{c}_i$ the residual capacity of knapsack i), then assign j to i and set $j = j+1$; (2) otherwise, (a) reject the current item (by setting $j = j+1$), (b) decide that the current knapsack is full (by setting $i = i+1$), and (c) waste (!) part of the capacity of the next knapsack (by setting $\overline{c}_i = c_i - (w_{j-1} - \overline{c}_{i-1})$). Clearly, this is a "stupid" algorithm, whose average performance can be immediately improved by eliminating step (c). The worst-case performance ratio, however, is not improved, since for the tightness counter-example above we still have $z^g = k + m + 1$. Trying to further improve the algorithm, we could observe that, in case (2), it rejects an item which could fit into some other knapsack and "closes" a knapsack which could contain some more items. However, if we restrict our attention to $O(n)$ algorithms which only go forward, i.e. never decrease the value of j or i, then by performing, in case (2), only step (a) or only step (b), the worst-case performance is not improved. If just j is increased, then for the same tightness counter-example we continue to have $z^g = k+m+1$. If just i is increased, then for the series of instances with $n = m+3, c_1 = 2k$ $(k > 1), c_i = k$ for $i = 2, \ldots, m$, $p_1 = w_1 = p_2 = w_2 = k+1$ and $p_j = w_j = k$ for $j = 3, \ldots, n$, we have $z = (m+1)k$ and $z^g = k+1$.

Other heuristic algorithms which, for example, for each item j perform a search among the knapsacks, are considered in Section 6.6.

6.4 EXACT ALGORITHMS

The optimal solution of MKP is usually obtained through branch-and-bound. Dynamic programming is in fact impractical for problems of this kind, both as regards computing times and storage requirements. (Note in addition that this approach would, for a strongly NP-hard problem, produce a strictly exponential time complexity.)

Algorithms for MKP are generally oriented either to the case of low values of the ratio n/m or to the case of high values of this ratio. Algorithms for the first class (which has applications, for example, when m liquids, which cannot be mixed, have to be loaded into n tanks) have been presented by Neebe and Dannenbring (1977) and by Christofides, Mingozzi and Toth (1979). In the following we will review algorithms for the second class, which has been more extensively studied.

6.4.1 Branch-and-bound algorithms

Hung and Fisk (1978) proposed a depth-first branch-and-bound algorithm in which successive levels of the branch-decision tree are constructed by selecting an item and assigning it to each knapsack in turn. When all the knapsacks have been considered, the item is assigned to a dummy knapsack, $m+1$, implying its exclusion from the current solution. Two implementations have been obtained by computing the upper bound associated with each node as the solution of the Lagrangian relaxation, or the surrogate relaxation of the current problem. The corresponding multipliers, $\overline{\lambda}$ and $\overline{\pi}$, have been determined as the optimal dual variables associated with constraints (6.3) and (6.2), respectively, in the continuous relaxation of the current problem (see Section 6.2.2). The choice of the item to be selected at each level of the decision-tree depends on the relaxation employed: in the Lagrangian case, the algorithm selects the item which, in the solution of the relaxed problem, has been inserted in the highest number of knapsacks; in the surrogate case, the item of lowest index is selected from among those which are still unassigned (i.e., at each level j, item j is selected). The items are sorted according to (6.9), the knapsacks so that

$$c_1 \geq c_2 \geq \ldots \geq c_m.$$

Once the branching item has been selected, it is assigned to knapsacks according to the increasing order of their indices. Figure 6.1 shows the decision nodes generated, when $m = 4$, for branching item j.

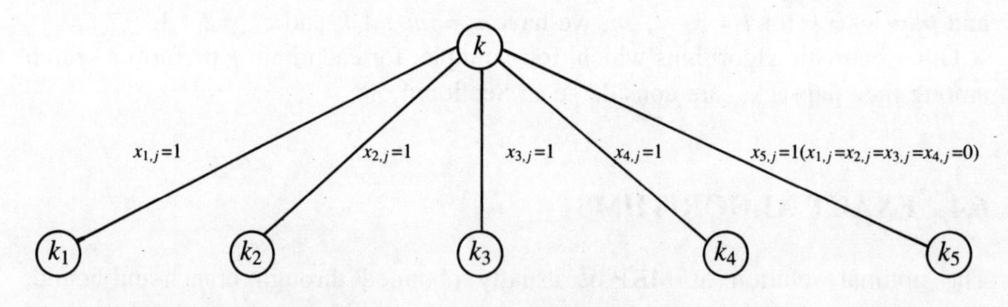

Figure 6.1 Branching strategy for the algorithms of Hung and Fisk (1978)

Martello and Toth (1980a) proposed a depth-first branch-and-bound algorithm using a different branching strategy based on the solution, at each decision node, of the current problem with constraints (6.3) dropped out. From (6.18)–(6.20) it is clear that the resulting relaxed problem coincides with a Lagrangian relaxation with

$\lambda_j = 0$ for $j = 1, \ldots, n$. In the following, this is denoted by $L(MKP, 0)$. For the instance of Example 6.1, we obtain: $(x_{1,j}) = (0, 1, 0, 0, 0, 1)$, $z_1 = 155$, $(x_{2,j}) = (1, 0, 0, 1, 0, 1)$, $z_2 = 195$, so $z(L(MKP, 0)) = 350$. In this case $L(MKP, 0)$ gives a better result than $L(MKP, \overline{\lambda})$. It is not difficult, however, to construct examples for which $z(L(MKP, \overline{\lambda})) < z(L(MKP, 0))$, i.e. neither of the two choices for λ dominates the other. In general, one can expect that the choice $\lambda = \overline{\lambda}$ produces tighter bounds. However, use of $\lambda = (0, \ldots, 0)$ in a branch-and-bound algorithm gives two important advantages:

(a) if no item is assigned to more than one knapsack in the solution of $L(MKP, 0)$, a *feasible and optimal* solution of the current problem has been found, and a backtracking can be immediately performed. If the same happens for $L(MKP, \lambda)$, with $\lambda \neq (0, \ldots, 0)$, the solution is just *feasible* (it is also optimal only when the corresponding value of the original objective function (6.1) equals $z(L(MKP, \lambda))$);

(b) since $(\tilde{p}_j)$ does not change from one level to another, the computation of the upper bounds associated with the decision nodes involves the solution of a lesser number of different 0-1 single knapsack problems.

The strategy adopted in Martello and Toth (1980a) is to select an item for branching which, in solution $(\hat{x}_{ij})$ to the current $L(MKP, 0)$, is inserted into $\overline{m} > 1$ knapsacks (namely, that having the maximum value of $(p_j/w_j) \sum_{i \in M} \hat{x}_{ij}$ is selected). $\overline{m}$ nodes are then generated, by assigning the item in turn to $\overline{m} - 1$ of such knapsacks and by excluding it from these. Suppose that, in the case of Figure 6.1, we have, for the selected item j, $\hat{x}_{1,j} = \hat{x}_{2,j} = \hat{x}_{3,j} = 1$ and $\hat{x}_{4,j} = 0$. Figure 6.2 shows the decision nodes generated.

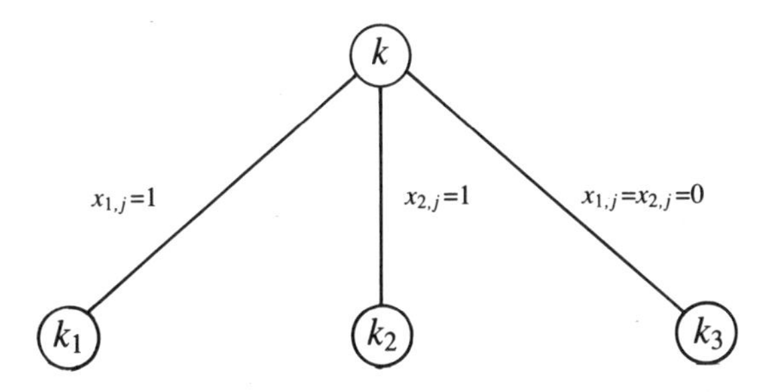

Figure 6.2 Branching strategy for the Martello and Toth (1980a) algorithm

In order to compute the upper bound associated with node k_1 it is sufficient to solve two single knapsack problems: the former for knapsack 2 with condition $x_{2,j} = 0$, the latter for knapsack 3 with condition $x_{3,j} = 0$ (the solutions for knapsacks 1 and 4 are unchanged with respect to those corresponding to the father node k). The upper bound associated with node k_2 can now be computed by solving only the single knapsack problem for knapsack 1 with condition $x_{1,j} = 0$, the solution of knapsack 3 with condition $x_{3,j} = 0$ having already been computed. Obviously, no single knapsack need now be solved to compute the upper bound associated with node k_3. In general, it is clear that $\overline{m}-1$ single knapsacks have to be solved for the first node considered, then one for the second node and none for the remaining $\overline{m}-2$ nodes. Hence, in the worst case ($\overline{m} = m$), only m single knapsack problems have to be solved in order to compute the upper bounds associated with the nodes which each node generates.

In addition we can compute, without solving any further single knapsack problem, the upper bound corresponding to the exclusion of the branching item j from all the $\overline{m}$ knapsacks considered: if this bound is not greater than the best solution so far, it is possible to associate a stronger condition with the branch leading to the $\overline{m}$th node by assigning the object to the $\overline{m}$th knapsack without changing the corresponding upper bound. In the example of Figure 6.2, condition $x_{1,j} = x_{2,j} = 0$ would be replaced by $x_{3,j} = 1$.

A further advantage of this strategy is that, since all the upper bounds associated with the $\overline{m}$ generated nodes are easily computed, the nodes can be explored in decreasing order of their upper bound values.

6.4.2 The "bound-and-bound" method

In Martello and Toth (1981a), MKP has been solved by introducing a modification of the branch-and-bound technique, based on the computation at each decision node not only of an upper bound, but also of a lower bound for the current problem. The method, which has been called *bound-and-bound*, can be used, in principle, to solve any integer linear program. In the next section we describe the resulting algorithm for MKP. Here we introduce the method for the general *0-1 Linear Programming Problem* (ZOLP)

$$\text{maximize} \quad \sum_{j \in N} p_j x_j$$

$$\text{subject to} \quad \sum_{j \in N} a_{ij} x_j \leq b_i, \qquad i \in M,$$

$$x_j = 0 \text{ or } 1, \qquad j \in N.$$

Let us suppose, for the sake of simplicity, that all coefficients are non-negative. We define a *partial solution* S as a set, represented as a stack, containing the indices of those variables whose value is fixed: an index in S is *labelled* if the value of the corresponding variable is fixed to 0, *unlabelled* if it is fixed to 1. The *current problem* induced by S, $ZOLP(S)$, is ZOLP with the additional constraints $x_j = 0$ ($j \in S, j$ labelled), $x_j = 1$ ($j \in S, j$ unlabelled).

Let $U(S)$ be any *upper bound* on $z(ZOLP(S))$. Let H be a heuristic procedure which, when applied to $ZOLP(S)$, has the following properties:

(i) a feasible solution $(\tilde{x}_j)$ is always found, if one exists;

(ii) this solution is maximal, in the sense that no $\tilde{x}_j$ having value 0 can be set to 1 without violating the constraints.

The value of the solution produced by H, $L(S) = \sum_{j \in N} p_j \tilde{x}_j$, is obviously a *lower bound* on $z(ZOLP(S))$.

A bound-and-bound algorithm for the optimal solution of ZOLP works as follows.

```
procedure BOUND_AND_BOUND:
input: N, M, (p_j), (a_ij), (b_i);
output: z, (x_j);
begin
1. [initialize]
   S := Ø;
   z := -∞ ;
2. [heuristic]
   apply heuristic procedure H to ZOLP(S);
   if ZOLP(S) has no feasible solution then go to 4;
   if L(S) > z then
        begin
             z := L(S);
             for each j ∈ N do x_j := x̃_j;
             if z = U(S) then go to 4
        end;
3. [define a new current solution]
   let j be the first index in N\S such that x̃_j = 1;
   if no such j then go to 4 ;
   push j (unlabelled) on S;
   if U(S) > z then go to 3;
4. [backtrack]
   while S ≠ Ø do
        begin
             let j be the index on top of S;
             if j is labelled then pop j from S;
             else
                  begin
                       label j;
                       if U(S) > z then go to 2 else go to 4
                  end
        end
end.
```

The main conceptual difference between this approach and a standard depth-first branch-and-bound one is that the branching phase is here performed by updating the partial solution through the heuristic solution determining the current lower bound. This gives two advantages:

(a) For all S for which $L(S) = U(S)$, $(\tilde{x}_j)$ is obviously an optimal solution to $ZOLP(S)$, so it is possible to avoid exploration of the decision nodes descending from the current one;

(b) For all S for which $L(S) < U(S)$, S is updated through the heuristic solution previously found by procedure H, so the resulting partial solution can generally be expected to be better than that which would be obtained by a series of forward steps, each fixing a variable independently of the following ones.

On the other hand, in case (b) it is possible that the computational effort spent to obtain $L(S)$ through H may be partially useless: this happens when, after few iterations of Step 3, condition $U(S) \leq z$ holds.

In general, the bound and bound approach is suitable for problems having the following properties:

(i) a "fast" heuristic procedure producing "good" lower bounds can be found;

(ii) the relaxation technique utilized to obtain the upper bounds leads to solutions whose feasibility for the current problem is difficult to check or is seldom verified.

6.4.3 A bound-and-bound algorithm

Martello and Toth (1981a) have derived from the previous framework an algorithm for MKP which consists of an enumerative scheme where each node of the decision-tree generates two branches either by assigning an item j to a knapsack i or by excluding j from i. For the sake of clarity, we give a description close to that of the general algorithm of the previous section, although this is not the most suitable for effective implementation. Stack S_k $(k = 1, \dots, m)$ contains those items that are currently assigned to knapsack k or excluded from it.

Let $S = \{S_1, \dots, S_m\}$. At each iteration, i denotes the current knapsack and the algorithm inserts in i the next item j selected, for knapsack i, by the current heuristic solution. Only when no further item can be inserted in i is knapsack $i+1$ considered. Hence, at any iteration, knapsacks $1, \dots, i-1$ are completely loaded, knapsack i is partially loaded and knapsacks $i+1, \dots, m$ are empty.

Upper bounds $U = U(S)$ are computed, through surrogate relaxation, by procedure UPPER. Lower bounds $L = L(S)$ and the corresponding heuristic solutions $\tilde{x}$ are computed by procedure LOWER, which finds an optimal solution for the current knapsack, then excludes the items inserted in it and finds an optimal solution for the next knapsack, and so on. For both procedures, on input i is the current knapsack and $(\hat{x}_{kj})$ $(k = 1, \dots, i;\ j = 1, \dots, n)$ contains the current solution.

```
procedure UPPER:
input: n, m, (p_j), (w_j), (c_k), (x̂_kj), (S_k), i;
output: U;
begin
    c̄ := (c_i − Σ_{j∈S_i} w_j x̂_ij) + Σ_{k=i+1}^{m} c_k;
    N̄ := {j : x̂_kj = 0 for k = 1, ..., i};
    determine the optimal solution value z̄ of the 0-1 single knapsack problem
      defined by the items in N̄ and by capacity c̄;
    U := Σ_{k=1}^{i} Σ_{j∈S_k} p_j x̂_kj + z̄
end.

procedure LOWER:
input: n, m, (p_j), (w_j), (c_k), (x̂_kj), (S_k), i;
output: L, (x̃_kj);
begin
    L := Σ_{k=1}^{i} Σ_{j∈S_k} p_j x̂_kj;
    N' := {j : x̂_kj = 0 for k = 1, ..., i};
    N̄ := N'\S_i;
    c̄ := c_i − Σ_{j∈S_i} w_j x̂_ij;
    k := i;
    repeat
        determine the optimal solution value z̄ of the 0-1 single knapsack problem
          defined by the items in N̄ and by capacity c̄, and store the solution
          vector in row k of x̃;
        L := L + z̄;
        N' := N'\{j : x̃_kj = 1};
        N̄ := N';
        k := k + 1;
        c̄ := c_k
    until k > m
end.
```

The bound-and-bound algorithm for MKP follows. Note that no current solution is defined (hence no backtracking is performed) for knapsack m since, given $\hat{x}_{kj}$ for $k = 1, \ldots, m-1$, the solution produced by LOWER for knapsack m is optimal. It follows that it is convenient to sort the knapsacks so that

$$c_1 \leq c_2 \leq \ldots \leq c_m.$$

The items are assumed to be sorted according to (6.9).

```
procedure MTM:
input: n, m, (p_j), (w_j), (c_i);
output: z, (x_ij);
begin
1. [initialize]
   for k := 1 to m do S_k := ∅;
```

```
   for k := 1 to m do for j := 1 to n do x̂_kj := 0;
   z := 0;
   i := 1;
   call UPPER yielding U;
   UB := U;
2. [heuristic]
   call LOWER yielding L and x̃;
   if L > z then
      begin
         z := L;
         for k := 1 to m do for j := 1 to n do x_kj := x̂_kj;
         for k := i to m do for j := 1 to n do
            if x̃_kj = 1 then x_kj := 1;
         if z = UB then return ;
         if z = U then go to 4
      end;
3. [define a new current solution]
   repeat
      I := {l : x̃_il = 1};
      while I ≠ Ø do
         begin
            let j = min{l : l ∈ I};
            I := I\{j};
            push j on S_i;
            x̂_ij := 1;
            call UPPER yielding U;
            if U ≤ z then go to 4
         end;
      i := i + 1
   until i = m;
   i := m − 1;
4. [backtrack]
   repeat
      while S_i≠Ø do
         begin
            let j be the item on top of S_i;
            if x̂_ij = 0 then pop j from S_i;
            else
               begin
                  x̂_ij = 0;
                  call UPPER yielding U;
                  if U > z then go to 2
               end
         end;
      i := i − 1
   until i = 0
end.
```

The Fortran implementation of procedure MTM (also presented in Martello and Toth (1985b)) is included in the present volume. With respect to the above description, it also includes a technique for the parametric computation of upper bounds U. In procedures UPPER and LOWER, the 0-1 single knapsack problems are solved through procedure MT1 of Section 2.5.2. (At each execution, the items are already sorted according to (6.9), so there would be no advantage in using procedure MT2 of Section 2.9.3.)

Example 6.2

Consider the instance of MKP defined by

$n = 10;$

$m = 2;$

$(p_j) = (78, 35, 89, 36, 94, 75, 74, 79, 80, 16);$

$(w_j) = (18, 9, 23, 20, 59, 61, 70, 75, 76, 30);$

$(c_i) = (103, 156).$

Applying procedure MTM, we obtain the branch decision-tree of Figure 6.3. At the nodes, $\hat{z}$ gives the current solution value, $(\hat{c}_i)$ the current residual capacities.

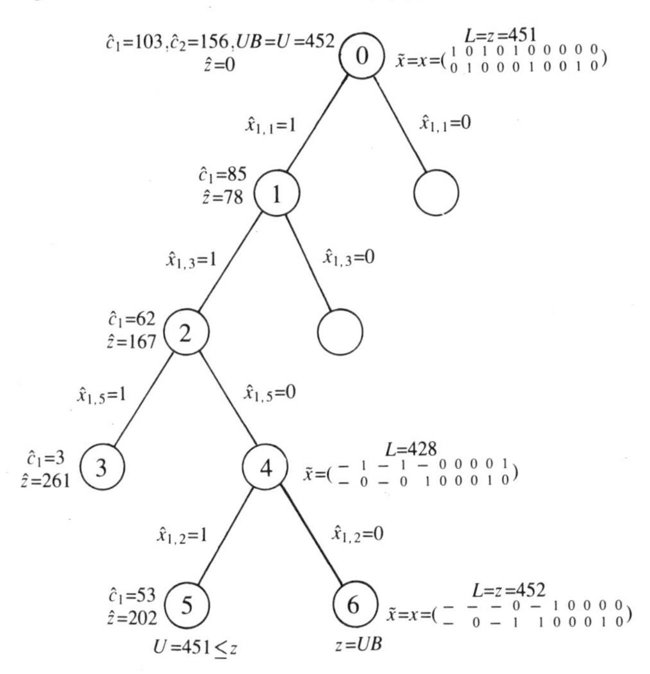

Figure 6.3 Decision-tree of procedure MTM for Example 6.2

The value of U is not given for the nodes for which the parametric computation was able to ensure that its value was the same as for the father node. The optimal solution is

$$(x_{ij}) = \begin{pmatrix} 1 & 0 & 1 & 0 & 0 & 1 & 0 & 0 & 0 & 0 \\ 0 & 0 & 0 & 1 & 1 & 0 & 0 & 0 & 1 & 0 \end{pmatrix},$$

$$z = 452. \square$$

A modified version of procedure MTM, in which dominance criteria among nodes of the decision-tree are applied, has been proposed by Fischetti and Toth (1988). Its performance is experimentally better for low values of the ratio n/m.

6.5 REDUCTION ALGORITHMS

The size of an instance of MKP can be reduced, as for the 0-1 knapsack problem (Section 2.7), by determining two sets, $J1$ and $J0$, containing those items which, respectively, must be and cannot be in an optimal solution. In this case, however, only $J0$ allows one to reduce the size of the problem by eliminating the corresponding items, while $J1$ cannot specify in which knapsack the items must be inserted, so it only gives information which can be imbedded in an implicit enumeration algorithm.

Ingargiola and Korsh (1975) presented a specific reduction procedure, based on dominance between items. Let $j\mathrm{D}k$ indicate that item j *dominates* item k, in the sense that, for any feasible solution that includes k but excludes j, there is a better feasible solution that includes j and excludes k. Consequently, if we can determine, for $j = 1, \dots, n$, a set D_j of items dominated by j, we can exclude all of them from the solution as soon as item j is excluded. If the items are sorted according to (6.9), D_j obviously contains all items $k > j$ such that $w_k \geq w_j$ and $p_k \leq p_j$, plus other items which can be determined as follows.

procedure IKRM:
input: $n, (p_k), (w_k), j$;
output: D_j;
begin
  $D_j := \{k : k > j, \ w_k \geq w_j$ and $p_k \leq p_j\}$;
  repeat
    $d := |D_j|$;
    for each $k \in \{l : p_l/w_l \leq p_j/w_j\} \backslash (D_j \cup \{j\})$ **do**
      if $\exists A \subseteq D_j : w_j + \sum_{a \in A} w_a \leq w_k$ and
        $p_j + \sum_{a \in A} p_a \geq p_k$
      then $D_j := D_j \cup \{k\}$
  until $|D_j| = d$
end.

The items added to D_j in the repeat-until loop are dominated by j since, for any solution that includes k but excludes j and, hence, all $a \in A$, there is a better solution that includes $\{j\} \cup A$ and excludes k. Once sets D_j $(j = 1, \dots, n)$ have been determined, if a feasible solution of value, say, $\overline{z}$ is known, a set of items which must be in an optimal solution is

$$J1 = \left\{ j : \sum_{k \in N \setminus (\{j\} \cup D_j)} p_k \leq \overline{z} \right\},$$

since the exclusion of any item $j \in J1$, and hence of all items in D_j, would not leave enough items to obtain a better solution. Observe now that, for any item k, set

$$I_k = \{ j : j \, \mathrm{D} k, j \notin J1 \}$$

contains items which must be included in any solution including k. Hence a set of items which must be excluded from an optimal solution is

$$J0 = \left\{ k : w_k + \sum_{j \in I_k} w_j > \sum_{i \in M} c_i - \sum_{j \in J1} w_j \right\}.$$

The time complexity of IKRM is $O(n^2 \varphi(n))$, where $\varphi(n)$ is the time required for the search of a suitable subset $A \subseteq D_j$. Exactly performing this search, however, requires exponential time, so a heuristic search should be used to obtain a polynomial algorithm. In any case, the overall time complexity for determining $J1$ and $J0$ is $O(n^3 \varphi(n))$, so the method can be useful only for low values of n or for very difficult problems.

6.6 APPROXIMATE ALGORITHMS

6.6.1 On the existence of approximation schemes

Let P be a maximization problem whose solution values are all positive integers. Let *length*(I) and *max*(I) denote, for any instance $I \in P$, the number of symbols required for encoding I and the magnitude of the largest number in I, respectively. Let $z(I)$ denote the optimal solution value for I. Then

Theorem 6.4 (Garey and Johnson, 1978) *If P is NP-hard in the strong sense and there exists a two-variable polynomial q such that, for any instance $I \in P$,*

$$z(I) < q(length(I), max(I)),$$

then P cannot be solved by a fully polynomial time approximation scheme unless $\mathcal{P} = \mathcal{NP}$.

Proof. Suppose such a scheme exists. By prefixing $1/\varepsilon = q(length(I), max(I))$, it would produce, in time polynomial in *length*(I) and $1/\varepsilon$ (hence in pseudo-polynomial time) a solution of value $z^h(I)$ satisfying $(z(I) - z^h(I))/z^h(I) \leq \varepsilon < 1/z(I)$, i.e. $z(I) - z^h(I) < 1$, hence optimal. But this is impossible, P being NP-hard in the strong sense. □ (The analogous result for minimization problems also holds.)

Theorem 6.4 rules out the existence of a *fully polynomial-time approximation scheme* for MKP, since the problem is NP-hard in the strong sense (see Section 1.3) and its solution value satisfies $z < n \max_j \{p_j\}+1$. Note that the same consideration applies to MKP in minimization form (defined by *minimize* (6.1), subject to: (6.2) with $\leq$ replaced by $\geq$, (6.3) and (6.4)), since its solution value satisfies $z > \min_j \{p_j\} - 1$.

As for the existence of a *polynomial-time approximation scheme*, the following general property can be used:

Theorem 6.5 (Garey and Johnson, 1979) *Let P be a minimization (resp. maximization) problem whose solution values are all positive integers and suppose that, for some fixed positive integer k, the decision problem "Given $I \in P$, is $z(I) \leq k$ (resp. $z(I) \geq k$) ?" is NP-complete. Then, if $\mathcal{P} \neq \mathcal{NP}$, no polynomial-time algorithm for P can produce a solution of value $z^h(I)$ satisfying*

$$\frac{z^h(I)}{z(I)} < 1 + \frac{1}{k} \quad \left(\text{resp.} \quad \frac{z(I)}{z^h(I)} < 1 + \frac{1}{k}\right)$$

and P cannot be solved by a polynomial-time approximation scheme.

Proof. We prove the thesis for the minimization case. Suppose such an algorithm exists. If $z^h(I) \leq k$ then, trivially, $z(I) \leq k$. Otherwise, $z^h(I) \geq k + 1$, so $z(I) > z^h(I)k/(k+1) \geq k$. Hence a contradiction, since the algorithm would solve an NP-complete problem in polynomial time. (The proof for the maximization case is almost identical.) □

We can use Theorem 6.5 to exclude the existence of a polynomial-time approximation scheme for MKP in minimization form. We use the value $k = 1$. Given any instance $(w_1, \dots, w_n)$ of PARTITION (see Section 1.3), define an instance of MKP in minimization form having $p_1 = 1, p_2 = \dots = p_n = 0$, an additional item with $p_{n+1} = 2$ and $w_{n+1} = \sum_{j=1}^{n} w_j$, and two knapsacks with $c_1 = c_2 = \frac{1}{2}\sum_{j=1}^{n} w_j$. Deciding whether the solution value is no greater than 1 is NP-complete, since the answer is *yes* if and only if the answer for the instance of PARTITION is *yes*.

For MKP in maximization form, instead, no proof is known, to our knowledge, for ruling out the existence of a polynomial-time approximation scheme, although no such scheme is known.

6.6.2 Polynomial-time approximation algorithms

In Section 6.3 we have examined the worst-case performance of an $O(n)$ greedy algorithm for MKP. In Section 6.4.3 we have introduced an approximate algorithm (LOWER) requiring exact solution of m single knapsack problems, hence, in the worst case, a non-polynomial running time. A different non-polynomial heuristic approach has been proposed by Fisk and Hung (1979), based on the exact solution of the surrogate relaxation, $S(MKP)$, of the problem. Let X_S denote the subset of items producing $z(S(MKP))$. The algorithm considers the items of X_S in decreasing order of weight, and tries to insert each item in a randomly selected knapsack or, if it does not fit, in any of the remaining knapsacks. When an item cannot be inserted in any knapsack, for each pair of knapsacks it attempts exchanges between items (one for one, then two for one, then one for two) until an exchange is found which fully utilizes the available space in one of the knapsacks. If all the items of X_S are inserted, an optimal solution is found; otherwise, the current (suboptimal) feasible solution can be improved by inserting in the knapsacks, in a greedy way, as many items of $N \backslash X_S$ as possible.

Martello and Toth (1981b) proposed a polynomial-time approximate algorithm which works as follows. The items are sorted according to (6.9), and the knapsacks so that

$$c_1 \leq c_2 \leq \ldots \leq c_m. \tag{6.29}$$

An initial feasible solution is determined by applying the greedy algorithm (Section 2.4) to the first knapsack, then to the second one by using only the remaining items, and so on. This is obtained by calling m times the following procedure, giving the capacity $\overline{c}_i = c_i$ of the current knapsack and the current solution, of value z, stored, for $j = 1, \ldots, n$, in

$$y_j = \begin{cases} 0 \text{ if item } j \text{ is currently unassigned;} \\ \text{index of the knapsack it is assigned to, otherwise.} \end{cases}$$

procedure GREEDYS:
input: $n, (p_j), (w_j), z, (y_j), i, \overline{c}_i$;
output: $z, (y_j)$;
begin
  for $j := 1$ **to** n **do**
    if $y_j = 0$ and $w_j \leq \overline{c}_i$ **then**
      begin
        $y_j := i$;
        $\overline{c}_i := \overline{c}_i - w_j$;
        $z := z + p_j$
      end
end.

After GREEDYS has been called m times, the algorithm improves on the solution through local exchanges. First, it considers all pairs of items assigned to different knapsacks and, if possible, interchanges them should the insertion of a new item into the solution be allowed. When all pairs have been considered, the algorithm tries to exclude in turn each item currently in the solution and to replace it with one or more items not in the solution so that the total profit is increased.

Computational experiments (Martello and Toth, 1981b) indicated that the exchanges tend to be much more effective when, in the current solution, each knapsack contains items having dissimilar profit per unit weight. This, however, is not the case for the initial solution determined with GREEDYS. In fact, for the first knapsacks, the best items are initially inserted and, after the critical item has been encountered, generally other "good" items of smaller weight are selected. It follows that, for the last knapsacks, we can expect that only "bad" items are available. Hence, the exchange phases are preceded by a rearrangement of the initial solution. This is obtained by removing from the knapsacks all the items currently in the solution, and reconsidering them according to *increasing* profit per unit weight, by trying to assign each item to the next knapsack, in a cyclic manner. (In this way the items with small weight are considered when the residual capacities are small.)

The resulting procedure follows. It is assumed that items and knapsacks are sorted according to (6.9) and (6.29).

procedure MTHM:
input: $n, m, (p_j), (w_j), (c_i)$;
output: $z, (y_j)$;
begin
1. [initial solution]
    $z := 0$;
    for $j := 1$ **to** n **do** $y_j := 0$;
    for $i := 1$ **to** m **do**
        begin
            $\overline{c}_i := c_i$;
            call GREEDYS
        end;
2. [rearrangement]
    $z := 0$;
    for $i := 1$ **to** m **do** $\overline{c}_i := c_i$;
    $i := 1$;
    for $j := n$ **to** 1 **step**-1 **do if** $y_j > 0$ **then**
        begin
            let l be the first index in $\{i, \dots, m\} \cup \{1, \dots, i-1\}$ such that $w_j \le \overline{c}_l$;
            if no such l **then** $y_j := 0$ **else**
                begin
                    $y_j := l$;
                    $\overline{c}_l := \overline{c}_l - w_j$;
                    $z := z + p_j$;
                    if $l < m$ **then** $i := l + 1$ **else** $i := 1$

```
                end
        end;
    for i := 1 to m do call GREEDYS;
 3. [first improvement]
    for j := 1 to n do if y_j > 0 then
        for k := j + 1 to n do if 0 < y_k ≠ y_j then
            begin
                h := arg max{w_j, w_k};
                l := arg min{w_j, w_k};
                d := w_h − w_l;
                if d ≤ c̄_{y_l} and c̄_{y_h} + d ≥ min{w_u : y_u = 0} then
                    begin
                        t := arg max{p_u : y_u = 0 and w_u ≤ c̄_{y_h} + d};
                        c̄_{y_h} := c̄_{y_h} + d − w_t;
                        c̄_{y_l} := c̄_{y_l} − d;
                        y_t := y_h;
                        y_h := y_l;
                        y_l := y_t;
                        z := z + p_t
                    end
            end;
 4. [second improvement]
    for j := n to 1 step-1 do if y_j > 0 then
        begin
            c̄ := c̄_{y_j} + w_j;
            Y := Ø;
            for k := 1 to n do
                if y_k = 0 and w_k ≤ c̄ then
                    begin
                        Y := Y ∪ {k};
                        c̄ := c̄ − w_k
                    end;
            if Σ_{k∈Y} p_k > p_j then
                begin
                    for each k ∈ Y do y_k := y_j;
                    c̄_{y_j} := c̄;
                    y_j := 0;
                    z := z + Σ_{k∈Y} p_k − p_j
                end
        end
end.
```

No step of MTHM requires more than $O(n^2)$ time. This is obvious for Steps 1 and 2 (since GREEDYS takes $O(n)$ time) and for Step 4. As for Step 3, it is enough to observe that the updating of $\min\{w_u : y_u = 0\}$ and the search for t (in the inner loop) are executed only when a new item enters the solution, hence $O(n)$ times in total.

The Fortran implementation of MTHM is included in the present volume. With respect to the above description: (a) at Step 1 it includes the possibility of using, for small-size problems, a more effective (and time consuming) way for determining the initial solution; (b) Step 3 incorporates additional tests to avoid the examination of hopeless pairs; (c) the execution of Step 4 is iterated until no further improvement is found. (More details can be found in Martello and Toth (1981b).)

Example 6.3

Consider the instance of MKP defined by

$n = 9$;

$m = 2$;

$(p_j) = (80, 20, 60, 40, 60, 60, 65, 25, 30)$;

$(w_j) = (40, 10, 40, 30, 50, 50, 55, 25, 40)$;

$(c_i) = (100, 150)$.

After Step 1 we have

$(y_j) = (1, 1, 1, 2, 2, 2, 0, 0, 0)$,

$z = 320$.

Step 2 changes (y_j) to

$(y_j) = (2, 1, 2, 1, 2, 1, 0, 0, 0)$, with $(\overline{c}_i) = (10, 20)$.

Step 3 interchanges items 1 and 4, and produces

$(y_j) = (1, 1, 2, 2, 2, 1, 0, 2, 0)$, with $(\overline{c}_i) = (0, 5)$,

$z = 345$.

Step 4 excludes item 5, and produces

$(y_j) = (1, 1, 2, 2, 0, 1, 2, 2, 0)$, with $(\overline{c}_i) = (0, 0)$,

$z = 350$,

which is the optimal solution. □

6.7 COMPUTATIONAL EXPERIMENTS

Tables 6.1 and 6.2 compare the Fortran IV implementations of the exact algorithms of the previous sections on randomly generated test problems, using *uncorrelated items* with

Table 6.1 Uncorrelated items; dissimilar capacities. CDC-Cyber 730 in seconds. Average times over 20 problems

m	n	HF	MT	MTM	IKRM + MTM
	25	0.221	0.143	0.076	0.119
2	50	0.694	0.278	0.112	0.333
	100	1.614	1.351	0.159	1.297
	200	6.981	7.182	0.223	6.551
	25	4.412	9.363	0.458	0.463
3	50	54.625	17.141	0.271	0.472
	100	—	—	0.327	1.542
	200	—	—	0.244	6.913
	25	time limit	time limit	1.027	0.921
4	50	—	—	0.952	1.102
	100	—	—	0.675	1.892
	200	—	—	0.518	7.084

Table 6.2 Uncorrelated items; similar capacities. CDC-Cyber 730 in seconds. Average times over 20 problems

m	n	HF	MT	MTM	IKRM + MTM
	25	0.280	0.141	0.191	0.215
2	50	0.671	0.473	0.329	0.490
	100	1.666	0.810	0.152	1.295
	200	6.109	4.991	0.313	6.733
	25	3.302	1.206	1.222	1.101
3	50	44.100	2.362	0.561	0.757
	100	—	6.101	0.428	1.622
	200	—	39.809	0.585	7.190
	25	13.712	6.341	3.690	3.351
4	50	time limit	26.100	12.508	9.516
	100	—	—	3.936	3.064
	200	—	—	9.313	7.412

p_j and w_j uniformly random in [10, 100],

and two classes of capacities: *dissimilar capacities*, having

$$c_i \text{ uniformly random in } \left[0, \left(0.5 \sum_{j=1}^{n} w_j - \sum_{k=1}^{i-1} c_k\right)\right] \quad \text{for } i = 1, \ldots, m-1,$$

and *similar capacities*, having

c_i uniformly random in $\left[0.4 \sum_{j=1}^{n} w_j / m, 0.6 \sum_{j=1}^{n} w_j / m \right]$ for $i = 1, \dots, m-1$.

For both classes, the capacity of the mth knapsack was set to

$$c_m = 0.5 \sum_{j=1}^{n} w_j - \sum_{i=1}^{m-1} c_i .$$

Whenever an instance did not satisfy conditions (6.5)–(6.8), a new instance was generated. The entries in the tables give average running times, expressed in seconds, comprehensive of the sorting times.

For each value of m and n, 20 instances were generated and solved on a CDC-Cyber 730. Each algorithm had a time limit of 300 seconds to solve the 80 instances generated for each value of m. When this limit was reached, we give the average time only if the number of solved instances was significant.

Tables 6.1 and 6.2 compare, on small-size problems, the branch-and-bound algorithms of Hung and Fisk (1978) and Martello and Toth (1980a) (Section 6.4.1) and the bound-and-bound algorithm MTM (Section 6.4.3). Three implementations of the Hung and Fisk (1978) algorithm are possible, according to the relaxation used (Lagrangian, surrogate, or a combination of the two). In addition, the algorithm can be run with or without previous application of the Ingargiola and Korsh (1975) reduction procedure IKRM (Section 6.5). Each entry in columns HF gives the lowest of the six average times obtained. Similarly, columns MT give the lowest of the four times obtained for the Martello and Toth (1980a) algorithm (Lagrangian or combination of Lagrangian and surrogate relaxation, with or without the application of IKRM). The last two columns refer to algorithm MTM, without and with the application of IKRM, respectively. For all the algorithms, the solution of the 0-1 single knapsack problems was obtained using algorithm MT1 of Section 2.5.2.

The tables show that MTM is the fastest method, and that use of the reduction procedure generally produces a considerable increase in the total computing time (except for very difficult problems). MT is generally faster than HF. The different capacity generations have little effect on HF and MT. For MTM, instead, problems with dissimilar capacities are considerably easier. This can be explained by observing that the algorithm generates no decision nodes for the last knapsack, so it is at an advantage when one of the capacities is much greater than the others. We used problems with dissimilar capacities to test MTM on larger instances.

Table 6.3 compares the exact algorithm MTM with the approximate algorithm MTHM. In addition, we analyse the behaviour of MTM when used to produce approximate solutions, by halting execution after B backtrackings (with B = 10 or 50). For each approximate algorithm we give, in brackets, the average percentage error. The table shows that the time required to find the exact solution increases much more steeply with m than with n and tends to become impractical for $m > 10$.

Table 6.3 Uncorrelated items; dissimilar capacities. CDC-Cyber 730 in seconds. Average times (average percentage errors) over 20 problems

m	n	MTM exact time	MTHM time (% error)	MTM (B = 10) time (% error)	MTM (B = 50) time (% error)
	50	0.082	0.013(0.170)	0.049(0.028)	0.070(0.004)
	100	0.129	0.031(0.147)	0.089(0.018)	0.127(0.000)
2	200	0.153	0.057(0.049)	0.143(0.000)	0.152(0.000)
	500	0.243	0.132(0.020)	0.242(0.000)	0.242(0.000)
	1 000	0.503	0.266(0.003)	0.502(0.000)	0.502(0.000)
	50	1.190	0.018(0.506)	0.157(0.344)	0.434(0.312)
	100	1.014	0.040(0.303)	0.268(0.076)	0.601(0.027)
5	200	1.178	0.074(0.148)	0.327(0.018)	0.687(0.012)
	500	0.862	0.186(0.031)	0.659(0.001)	0.705(0.001)
	1 000	1.576	0.391(0.016)	1.231(0.001)	1.576(0.000)
	50	3.852	0.035(0.832)	0.162(0.287)	0.477(0.211)
	100	7.610	0.057(0.437)	0.324(0.174)	0.950(0.092)
10	200	32.439	0.106(0.219)	0.659(0.060)	1.385(0.039)
	500	5.198	0.535(0.078)	1.760(0.009)	3.836(0.003)
	1 000	9.729	0.870(0.031)	3.846(0.003)	7.623(0.001)

When used as a heuristic, MTM gives solutions very close to the optimum; the running times are reasonable and increase slowly with n and m. MTHM is faster than MTM but its solutions are clearly worse.

Tables 6.4 and 6.5 show the behaviour of approximate algorithms (MTM halted after 10 backtrackings and MTHM) on very large-size instances. The Fisk and Hung (1979) algorithm is not considered, since extensive computational experiments (Martello and Toth, 1981b) showed that it is generally dominated by MTHM. All runs were executed on an HP 9000/840 with option "-o" for the Fortran compiler. We used the same capacity generations as in the previous tables. For all data generations, for $n \geq 5000$ the execution of MTHM was halted at the end of Step 3, so as to avoid the most time consuming phase (this is possible through an input parameter in the corresponding Fortran implementation).

Table 6.4 refers to uncorrelated items, obtained by generating

$$p_j \text{ and } w_j \text{ uniformly random in } [1, 1000].$$

The percentage errors were computed with respect to the optimal solution value for $m \leq 5$, with respect to the initial upper bound determined by MTM for larger values. With few exceptions in the case of very large problems, both algorithms require acceptable computing times. The approximation obtained is generally very good. The times of MTM (B = 10) are one order of magnitude larger than those of MTHM, but the errors produced are one order of magnitude smaller. Computational experiments on *weakly correlated* items (w_j uniformly random in [1, 1000], p_j uniformly random in $[w_j - 100, w_j + 100]$) gave similar results, both for computing times and percentage errors.

Table 6.4 Uncorrelated items. HP 9000/840 in seconds. Average times (average percentage errors) over 20 problems

		Dissimilar capacities		Similar capacities	
m	n	MTHM	MTM (B = 10)	MTHM	MTM (B = 10)
	200	0.266(0.0694)	0.131(0.0049)	0.277(0.0441)	0.157(0.0081)
	500	0.085(0.0208)	0.382(0.0006)	0.086(0.0197)	0.387(0.0011)
2	1000	0.177(0.0048)	0.877(0.0001)	0.173(0.0059)	0.728(0.0002)
	2000	0.359(0.0017)	1.354(0.0001)	0.392(0.0023)	1.638(0.0000)
	5000	0.806(0.0009)	3.716(0.0000)	0.802(0.0007)	3.346(0.0000)
	10000	1.730(0.0004)	4.962(0.0000)	1.691(0.0003)	5.250(0.0000)
	200	0.418(0.1796)	0.283(0.0235)	0.529(0.2152)	0.328(0.0275)
	500	0.104(0.0278)	0.942(0.0037)	0.109(0.0408)	1.022(0.0069)
5	1000	0.214(0.0105)	2.009(0.0014)	0.203(0.0146)	1.976(0.0012)
	2000	0.455(0.0038)	3.510(0.0003)	0.409(0.0048)	3.994(0.0003)
	5000	0.968(0.0010)	7.348(0.0000)	0.888(0.0011)	9.849(0.0000)
	10000	1.998(0.0004)	9.138(0.0000)	1.843(0.0005)	23.932(0.0000)
	200	0.064(0.1826)	0.500(0.0582)	0.052(0.3051)	0.046(0.1024)
	500	0.154(0.0344)	1.172(0.0094)	0.132(0.0762)	1.373(0.0135)
10	1000	0.300(0.0143)	2.517(0.0022)	0.262(0.0189)	2.561(0.0032)
	2000	0.685(0.0041)	6.608(0.0004)	0.531(0.0079)	7.030(0.0008)
	5000	1.273(0.0009)	8.502(0.0000)	1.143(0.0022)	14.127(0.0001)
	10000	2.527(0.0004)	15.773(0.0000)	2.294(0.0007)	45.760(0.0000)
	200	0.100(0.1994)	0.706(0.0865)	0.088(0.9004)	0.614(0.2619)
	500	0.245(0.0471)	1.671(0.0181)	0.198(0.1393)	1.783(0.0327)
20	1000	0.426(0.0136)	4.285(0.0051)	0.403(0.0448)	4.065(0.0075)
	2000	0.796(0.0059)	7.332(0.0012)	0.754(0.0113)	11.717(0.0016)
	5000	1.676(0.0015)	17.980(0.0002)	1.659(0.0028)	27.829(0.0002)
	10000	3.191(0.0005)	30.608(0.0000)	3.466(0.0010)	84.605(0.0000)
	200	0.188(0.2865)	1.218(0.1923)	0.179(2.4654)	0.995(1.1246)
	500	0.446(0.0752)	3.501(0.0477)	0.378(0.4732)	2.748(0.0808)
40	1000	0.910(0.0255)	7.575(0.0137)	0.696(0.1219)	6.049(0.0173)
	2000	1.411(0.0081)	12.689(0.0039)	1.289(0.0364)	13.608(0.0041)
	5000	3.085(0.0022)	27.718(0.0009)	2.761(0.0065)	44.538(0.0004)
	10000	5.733(0.0008)	37.310(0.0004)	5.364(0.0020)	124.637(0.0001)

Table 6.5 shows the behaviour of MTHM on *strongly correlated* items, obtained with

$$w_j \text{ uniformly random in } [1, 1000],$$

$$p_j = w_j + 100.$$

MTM was not run since it requires the exact solution of 0-1 single knapsack problems, which is practically impossible for this data generation (see Section 2.10.1). The percentage errors were computed with respect to an upper bound on the solution value of the surrogate relaxation of the problem (we used upper bound U_2 of Section 2.3.1). The computing times are slightly higher than for uncorrelated items; the percentage errors are higher for large values of n.

Table 6.5 Algorithm MTHM. Strongly correlated items. HP 9000/840 in seconds. Average times (average percentage errors) over 20 problems

m	n	Dissimilar capacities	Similar capacities
	200	0.124(0.0871)	0.114(0.0803)
	500	0.829(0.0422)	0.460(0.0278)
2	1 000	1.546(0.0157)	1.078(0.0138)
	2 000	5.333(0.0069)	7.498(0.0083)
	5 000	0.823(0.0236)	0.805(0.0191)
	10 000	1.618(0.0144)	1.571(0.0110)
	200	0.165(0.1085)	0.130(0.1061)
	500	0.683(0.0364)	0.373(0.0313)
5	1 000	1.832(0.0155)	1.214(0.0133)
	2 000	3.500(0.0072)	6.662(0.0076)
	5 000	1.068(0.0272)	0.917(0.0245)
	10 000	2.173(0.0142)	1.919(0.0097)
	200	0.158(0.1466)	0.091(0.1498)
	500	0.636(0.0383)	0.668(0.0443)
10	1 000	1.583(0.0167)	1.217(0.0132)
	2 000	9.943(0.0090)	7.862(0.0079)
	5 000	1.697(0.0278)	1.214(0.0255)
	10 000	3.246(0.0134)	2.507(0.0112)
	200	0.154(0.6698)	0.194(0.3539)
	500	0.491(0.0624)	0.480(0.0558)
20	1 000	1.172(0.0187)	1.833(0.0195)
	2 000	7.293(0.0091)	5.728(0.0082)
	5 000	2.624(0.0237)	1.802(0.0285)
	10 000	5.307(0.0096)	3.686(0.0179)
	200	0.249(4.2143)	0.446(2.3671)
	500	0.807(0.4680)	1.369(0.1365)
40	1 000	1.460(0.0491)	3.477(0.0302)
	2 000	6.481(0.0137)	9.776(0.0108)
	5 000	4.799(0.0241)	2.986(0.0432)
	10 000	9.695(0.0141)	6.031(0.0186)

7

Generalized assignment problem

7.1 INTRODUCTION

The *Generalized Assignment Problem* (GAP) can be described, using the terminology of knapsack problems, as follows. Given n *items* and m *knapsacks*, with

$$p_{ij} = \textit{profit} \text{ of item } j \text{ if assigned to knapsack } i,$$

$$w_{ij} = \textit{weight} \text{ of item } j \text{ if assigned to knapsack } i,$$

$$c_i = \textit{capacity} \text{ of knapsack } i,$$

assign each item to exactly one knapsack so as to maximize the total profit assigned, without assigning to any knapsack a total weight greater than its capacity, i.e.

$$\text{maximize} \quad z = \sum_{i=1}^{m} \sum_{j=1}^{n} p_{ij} x_{ij} \tag{7.1}$$

$$\text{subject to} \quad \sum_{j=1}^{n} w_{ij} x_{ij} \le c_i, \qquad i \in M = \{1, \dots, m\}, \tag{7.2}$$

$$\sum_{i=1}^{m} x_{ij} = 1, \qquad j \in N = \{1, \dots, n\}, \tag{7.3}$$

$$x_{ij} = 0 \text{ or } 1, \qquad i \in M, j \in N, \tag{7.4}$$

where

$$x_{ij} = \begin{cases} 1 & \text{if item } j \text{ is assigned to knapsack } i; \\ 0 & \text{otherwise.} \end{cases}$$

The problem is frequently described in the literature as that of optimally assigning n tasks to m processors (n jobs to m agents, and so on), given the profit p_{ij} and the amount of resource w_{ij} corresponding to the assignment of task j to processor i, and the total resource c_i available for each processor i.

The minimization version of the problem can also be encountered in the literature: by defining c_{ij} as the *cost* required to assign item j to knapsack i, MINGAP is

$$\text{minimize} \quad v = \sum_{i=1}^{m} \sum_{j=1}^{n} c_{ij} x_{ij} \tag{7.5}$$

$$\text{subject to} \qquad (7.2),\ (7.3),\ (7.4).$$

GAP and MINGAP are equivalent. Setting $p_{ij} = -c_{ij}$ (or $c_{ij} = -p_{ij}$) for all $i \in M$ and $j \in N$ immediately transforms one version into the other. If the numerical data are restricted to positive integers (as frequently occurs), the transformation can be obtained as follows. Given an instance of MINGAP, define any integer value t such that

$$t > \max_{i \in M,\, j \in N} \{c_{ij}\} \tag{7.6}$$

and set

$$p_{ij} = t - c_{ij} \qquad \text{for } i \in M,\ j \in N. \tag{7.7}$$

From (7.5) we then have

$$v = t \sum_{j=1}^{n} \sum_{i=1}^{m} x_{ij} - \sum_{i=1}^{m} \sum_{j=1}^{n} p_{ij} x_{ij},$$

where, from (7.3), the first term is independent of (x_{ij}). Hence the solution (x_{ij}) of GAP also solves MINGAP. The same method transforms any instance of GAP into an equivalent instance of MINGAP (by setting $c_{ij} = \hat{t} - p_{ij}$ for $i \in M,\ j \in N$, with $\hat{t} > \max_{i \in M,\, j \in N}\{p_{ij}\}$).

Because of constraints (7.3), an instance of the generalized assignment problem does not necessarily have a feasible solution. Moreover, even the feasibility question is NP-complete. In fact, given an instance $(w_1, \dots, w_n)$ of PARTITION (see Section 1.3), consider the instance of GAP (or MINGAP) having $m = 2$, $w_{1,j} = w_{2,j} = w_j$ and $p_{1,j} = p_{2,j} = 1$ for $j = 1, \dots, n$, and $c_1 = c_2 = \frac{1}{2} \sum_{j=1}^{n} w_j$. Deciding whether a feasible solution (of value n) to such instance exists is an NP-complete problem, since the answer is yes if and only if the answer to the instance of PARTITION is yes.

The following version of the problem (LEGAP), instead, always admits a feasible solution.

$$\text{maximize} \quad \hat{z} = \sum_{i=1}^{m} \sum_{j=1}^{n} \hat{p}_{ij} x_{ij} \tag{7.8}$$

$$\text{subject to} \qquad (7.2),\ (7.4) \text{ and}$$

$$\sum_{i=1}^{m} x_{ij} \le 1, \qquad j \in N. \tag{7.9}$$

LEGAP too is equivalent to GAP. Given any instance of LEGAP, an equivalent instance of GAP will have an additional knapsack of capacity $c_{m+1} = n$, with $p_{m+1,j} = 0$ and $w_{m+1,j} = 1$ for $j \in N$, while $p_{ij} = \hat{p}_{ij}$ for $i \in M$ and $j \in N$. (Knapsack $m + 1$ gives no extra profit and always allows a feasible solution, so $z = \hat{z}$.) Conversely, given any instance of GAP, we can define an integer constant q such that

$$q > \sum_{j \in N} \max_{i \in M} \{p_{ij}\},$$

and set

$$\hat{p}_{ij} = p_{ij} + q \qquad \text{for } i \in M, j \in N.$$

With these profits, any set of n items has a higher value than any set of $k < n$ items. Hence, by solving LEGAP we obtain the solution for GAP (of value $z = \hat{z} - nq$) if (7.3) is satisfied, or we know that the instance of GAP has no feasible solution if $\sum_{i=1}^{m} x_{ij} = 0$ for some j.

LEGAP is a generalization of the 0-1 multiple knapsack problem (Chapter 6), in which $p_{ij} = p_j$ and $w_{ij} = w_j$ for all $i \in M$ and $j \in N$ (i.e. the profit and weight of each item are independent of the knapsack it is assigned to). Lagrangian relaxations for LEGAP have been studied by Chalmet and Gelders (1977).

The best known special case of generalized assignment problem is the *Linear Min–Sum Assignment Problem* (or *Assignment Problem*), which is a MINGAP with $n = m$, $c_i = 1$ and $w_{ij} = 1$ for all $i \in M$ and $j \in N$ (so, because of (7.3), constraints (7.2) can be replaced by $\sum_{j=1}^{n} x_{ij} = 1$ for $i \in M$). The problem can be solved in $O(n^3)$ time through the classical Hungarian algorithm (Kuhn (1955), Lawler (1976); efficient Fortran codes can be found in Carpaneto, Martello and Toth (1988)). The assignment problem, however, is not used in general as a subproblem in algorithms for the generalized case.

Another special case arises when $w_{ij} = w_j$ for all $i \in M$ and $j \in N$. Implicit enumeration algorithms for this case have been presented by De Maio and Roveda (1971) and Srinivasan and Thompson (1973).

Facets of the GAP polytope have been studied by Gottlieb and Rao (1989a, 1989b).

We will suppose, as is usual, that the weights w_{ij} of any GAP instance are positive integers. Hence, without loss of generality, we will also assume that

$$p_{ij} \text{ and } c_i \text{ are positive integers,} \tag{7.10}$$

$$|\{i : w_{ij} \le c_i\}| \ge 1 \qquad \text{for } j \in N, \tag{7.11}$$

$$c_i \ge \min_{j \in N} \{w_{ij}\} \qquad \text{for } i \in M. \tag{7.12}$$

If assumption (7.10) is violated, (a) fractions can be handled by multiplying through by a proper factor; (b) knapsacks with $c_i \le 0$ can be eliminated; (c) for each item j having $\min_{i \in M}\{p_{ij}\} \le 0$, we can set $p_{ij} = p_{ij} + |\min_{i \in M}\{p_{ij}\}| + 1$

for $i \in M$ and subtract $|\min_{i \in M}\{p_{ij}\}| + 1$ from the resulting objective function value. As is the case for the 0-1 multiple knapsack problem, there is no easy way of transforming an instance so as to handle negative weights, but all our considerations easily extend to this case too. If an item violates assumption (7.11) then it cannot be assigned, so the GAP instance is infeasible. Knapsacks violating assumption (7.12) can be eliminated from the instance.

In Section 7.2 we introduce various types of relaxations. Exact and approximate algorithms are described in Sections 7.3 and 7.4, reduction procedures in Section 7.5. Section 7.6 presents the results of computational experiments.

7.2 RELAXATIONS AND UPPER BOUNDS

The continuous relaxation of GAP, $C(GAP)$, given by (7.1)–(7.3) and

$$x_{ij} \geq 0, \qquad i \in M, j \in N, \tag{7.13}$$

is rarely used in the literature since it does not exploit the structure of the problem and tends to give solutions a long way from feasibility.

7.2.1 Relaxation of the capacity constraints

Ross and Soland (1975) have proposed the following upper bound for GAP. First, constraints (7.2) are relaxed to

$$w_{ij}x_{ij} \leq c_i, \; i \in M, j \in N,$$

and the optimal solution $\hat{x}$ to the resulting problem is obtained by determining, for each $j \in N$,

$$i(j) = \arg\max \{p_{ij} : i \in M, \, w_{ij} \leq c_i\}$$

and setting $\hat{x}_{i(j),j} = 1$ and $\hat{x}_{ij} = 0$ for all $i \in M \backslash \{i(j)\}$. The resulting upper bound, of value

$$U_0 = \sum_{j=1}^{n} p_{i(j),j}, \tag{7.14}$$

is then improved as follows. Let

$$N_i = \{j \in N : \hat{x}_{ij} = 1\}, \qquad i \in M,$$

$$d_i = \sum_{j \in N_i} w_{ij} - c_i, \qquad i \in M,$$

$$M' = \{i \in M : d_i > 0\},$$

$$N' = \bigcup_{i \in M'} N_i.$$

Given a set S of numbers, we denote with $\max_2 S$ (resp. $\min_2 S$) the second maximum (resp. minimum) value in S, and with $\arg\max_2 S$ (resp. $\arg\min_2 S$) the corresponding index. Since M' is the set of those knapsacks for which the relaxed constraint (7.2) is violated,

$$q_j = p_{i(j),j} - \max_2\{p_{ij} : i \in M, w_{ij} \le c_i\}, \quad j \in N'$$

gives the minimum penalty that will be incurred if an item j currently assigned to a knapsack in M' is reassigned. Hence, for each $i \in M'$, a lower bound on the loss of profit to be paid in order to satisfy constraint (7.2) is given by the solution to the 0-1 single knapsack problem in minimization form (see Section 2.1), KP_i^1 ($i \in M'$), defined by

$$\text{minimize} \quad v_i = \sum_{j \in N_i} q_j y_{ij}$$

$$\text{subject to} \quad \sum_{j \in N_i} w_{ij} y_{ij} \ge d_i,$$

$$y_{ij} = 0 \text{ or } 1, \quad j \in N_i,$$

where $y_{ij} = 1$ if and only if item j is removed from knapsack i. The resulting Ross and Soland (1975) bound is thus

$$U_1 = U_0 - \sum_{i \in M'} v_i. \tag{7.15}$$

This bound can also be derived from the Lagrangian relaxation, $L(GAP, \lambda)$, of the problem, obtained by dualizing constraints (7.3) in much the same way as described in Section 6.2.2 for the 0-1 multiple knapsack problem. In this case too the relaxed problem,

$$\text{maximize} \quad \sum_{i=1}^{m} \sum_{j=1}^{n} p_{ij} x_{ij} - \sum_{j=1}^{n} \lambda_j \left(\sum_{i=1}^{m} x_{ij} - 1 \right)$$

$$\text{subject to} \quad (7.2), (7.4),$$

separates into m 0-1 single knapsack problems (KP_i^λ, $i = 1, \dots, m$) of the form

$$\text{maximize} \quad z_i = \sum_{j=1}^{n} \tilde{p}_{ij} x_{ij}$$

$$\text{subject to} \quad \sum_{j=1}^{n} w_{ij} x_{ij} \le c_i,$$

$$x_{ij} = 0 \text{ or } 1, \quad j \in N,$$

where $\tilde{p}_{ij} = p_{ij} - \lambda_j$, and its solution value is

$$z(L(GAP, \lambda)) = \sum_{i=1}^{m} z_i + \sum_{j=1}^{n} \lambda_j. \tag{7.16}$$

It is now easy to see that, by choosing for λ_j the value

$$\overline{\lambda}_j = \max_2\{p_{ij} : i \in M, \ w_{ij} \le c_i\}, \quad j \in N,$$

we have $z(L(GAP, \overline{\lambda})) = U_1$. In fact, by transforming each KP_i^1 into an equivalent maximization form (as described in Section 2.1), and noting that, in each KP_i^λ, $\tilde{p}_{ij} \le 0$ if $j \notin N_i$ and $w_{ij} \le c_i$, we have $v_i = \sum_{j \in N_i} q_j - z_i$ $(i \in M')$. Hence, from (7.14) and (7.15),

$$U_1 = \sum_{j \in N} p_{i(j),j} - \sum_{j \in N'} p_{i(j),j} + \sum_{j \in N'} \overline{\lambda}_j + \sum_{i \in M'} z_i;$$

observing that, for $i \notin M'$, by definition we have $\sum_{j \in N_i} w_{ij} \le c_i$, hence $z_i = \sum_{j \in N_i} \tilde{p}_{ij}$, the Lagrangian solution value (7.16) can be written as

$$\begin{aligned} z(L(GAP, \overline{\lambda})) &= \sum_{i \in M'} z_i + \sum_{i \in M \setminus M'} \sum_{j \in N_i} (p_{ij} - \overline{\lambda}_j) + \sum_{j \in N} \overline{\lambda}_j \\ &= \sum_{i \in M'} z_i + \sum_{j \in N \setminus N'} p_{i(j),j} - \sum_{j \in N \setminus N'} \overline{\lambda}_j + \sum_{j \in N} \overline{\lambda}_j \\ &= U_1. \end{aligned}$$

Example 7.1

Consider the instance of GAP defined by

$n = 7;$

$m = 2;$

$$(p_{ij}) = \begin{pmatrix} 6 & 9 & 4 & 2 & 10 & 3 & 6 \\ 4 & 8 & 9 & 1 & 7 & 5 & 4 \end{pmatrix};$$

$$(w_{ij}) = \begin{pmatrix} 4 & 1 & 2 & 1 & 4 & 3 & 8 \\ 9 & 9 & 8 & 1 & 3 & 8 & 7 \end{pmatrix}; \ (c_i) = \begin{pmatrix} 11 \\ 22 \end{pmatrix}.$$

The initial bound is $U_0 = 47$. Then we have

$N_1 = \{1, 2, 4, 5, 7\}$, $N_2 = \{3, 6\}$, $(d_i) = (7, -6)$;

$M' = \{1\}$, $N' = \{1, 2, 4, 5, 7\}$;

$q_1 = 2$, $q_2 = 1$, $q_4 = 1$, $q_5 = 3$, $q_7 = 2$.

Solving KP_1^1 we obtain

$v_1 = 2$, $(y_{1,j}) = (0, 0, -, 0, 0, -, 1)$,

so the resulting bound is

$U_1 = U_0 - v_1 = 45$. □

7.2.2 Relaxation of the semi-assignment constraints

Martello and Toth (1981c) have obtained an upper bound for GAP by removing constraints (7.3). It is immediate to see that the resulting relaxed problem coincides with $L(GAP, 0)$, hence it decomposes into a series of 0-1 single knapsack problems, KP_i^2 $(i \in M)$, of the form

$$\text{maximize} \quad z_i = \sum_{j=1}^{n} p_{ij} x_{ij}$$

$$\text{subject to} \quad \sum_{j=1}^{n} w_{ij} x_{ij} \leq c_i,$$

$$x_{ij} = 0 \text{ or } 1, \qquad j \in N.$$

In this case too, the resulting upper bound, of value

$$\overline{U}_0 = \sum_{i=1}^{m} z_i, \tag{7.17}$$

can be improved by computing a lower bound on the penalty to be paid in order to satisfy the violated constraints. Let

$$N^0 = \left\{ j \in N : \sum_{i \in M} x_{ij} = 0 \right\},$$

$$N^{>} = \left\{ j \in N : \sum_{i \in M} x_{ij} > 1 \right\}$$

be the sets of those items for which (7.3) is violated, and define

$$M^{>}(j) = \{ i \in M : x_{ij} = 1 \} \qquad \text{for all } j \in N^{>};$$

we can compute, using any of the methods of Sections 2.2–2.3,

$$u_{ij}^{0} = \text{upper bound on } z_i \text{ if } x_{ij} = 0, \quad j \in N^{>}, i \in M^{>}(j),$$

$$u_{ij}^{1} = \text{upper bound on } z_i \text{ if } x_{ij} = 1, \quad j \in N^{0}, i \in M$$

and determine, for each item $j \in N^0 \cup N^>$, a lower bound l_j on the penalty to be paid for satisfying (7.3):

$$l_j = \begin{cases} \min_{i \in M} \{ z_i - \min (z_i, u_{ij}^1) \} & \text{if } j \in N^0; \\ \sum_{i \in M^>(j)} (z_i - \min (z_i, u_{ij}^0)) & \\ \quad - \max_{i \in M^>(j)} \{ z_i - \min (z_i, u_{ij}^0) \} & \text{if } j \in N^>. \end{cases}$$

The improved upper bound is thus

$$U_2 = \overline{U}_0 - \max_{j \in N^0 \cup N^>} \{ l_j \}. \tag{7.18}$$

Example 7.2

Consider the instance of GAP defined by

$n = 5;$

$m = 2;$

$$(p_{ij}) = \begin{pmatrix} 7 & 3 & 3 & 8 & 7 \\ 5 & 3 & 8 & 4 & 1 \end{pmatrix};$$

$$(w_{ij}) = \begin{pmatrix} 8 & 2 & 8 & 9 & 1 \\ 2 & 2 & 6 & 4 & 4 \end{pmatrix};$$

$$(c_i) = \begin{pmatrix} 11 \\ 7 \end{pmatrix}.$$

The solutions to KP_1^2 and KP_2^2 are

$z_1 = 17, \ (x_{1,j}) = (1, 1, 0, 0, 1);$

$z_2 = \ 9, \ (x_{2,j}) = (1, 0, 0, 1, 0),$

so $\overline{U}_0 = 26$ and

$N^0 = \{3\}, N^> = \{1\}, M^>(1) = \{1, 2\}.$

We compute u_{ij}^0 and u_{ij}^1 through the Dantzig bound (Section 2.2.1), but, for u_{ij}^1, we skip those items k for which $w_{ik} > c_i - w_{ij}$. Hence

$$u_{1,1}^0 = 7 + 3 + \left\lfloor \frac{64}{9} \right\rfloor = 17;$$

$$u_{2,1}^0 = 3 + \left\lfloor \frac{40}{6} \right\rfloor = 9;$$

$$u_{1,3}^1 = 3 + (7 + 3 + \lfloor 0 \rfloor) = 13;$$

$$u_{2,3}^1 = 8.$$

It follows that $l_1 = 0$ and $l_3 = \min\{4, 1\} = 1$, so the resulting upper bound is

$U_2 = \overline{U}_0 - l_3 = 25.$

For this instance the Ross–Soland bound initially gives $U_0 = 33$, and, after the solution of KP_1^1, $U_1 = 31$. Hence $U_0 > U_1 > \overline{U}_0 > U_2$. On the other hand, computing the Martello–Toth bound for Example 7.1 gives $\overline{U}_0 = 54$, $l_2 = 2$, $l_3 = 1$, $l_5 = 5$, $l_6 = 1$, $l_7 = 2$, and $U_2 = 49$, i.e. $U_1 < U_0 < U_2 < \overline{U}_0$. Thus while, obviously, $U_0 \geq U_1$ and $\overline{U}_0 \geq U_2$, no dominance exists between the other pairs of these bounds. □

7.2.3 The multiplier adjustment method

Fisher, Jaikumar and Van Wassenhove (1986) have developed an upper bound, based on the Lagrangian relaxation $L(GAP, \lambda)$ and dominating the bound proposed by Ross and Soland (1975). Obviously, the continuous and integer solutions of a knapsack problem may differ; this implies (see Fisher (1981)) that, for the *optimal* Lagrangian multiplier λ^*,

$$z(L(GAP, \lambda^*)) \leq z(C(GAP));$$

there is no analytical way, however, to determine λ^*. One possibility is the classical subgradient optimization approach. The novelty of the Fisher–Jaikumar–Van Wassenhove bound consists of a new technique (*multiplier adjustment method*) for determining "good" multipliers. The method starts by setting

$$\lambda_j = \max{}_2 \{p_{ij} : i \in M, w_{ij} \leq c_i\}, \quad j \in N;$$

as shown in Section 7.2.1, the corresponding Lagrangian relaxation produces the value U_1 of the Ross–Soland bound. Note, in addition, that, with this choice, we have, for each $j \in N$, $\tilde{p}_{ij}$ $(= p_{ij} - \lambda_j) > 0$ for at most one $i \in M$, so there is an optimal Lagrangian solution for this λ which satisfies $\sum_{i=1}^{m} x_{ij} \leq 1$ for all $j \in N$. If some constraint (7.3) is not satisfied, it is, under certain conditions, possible to select a j^* for which $\sum_{i=1}^{m} x_{ij^*} = 0$ and decrease λ_{j^*} by an amount which ensures that in the new Lagrangian solution $\sum_{i=1}^{m} x_{ij^*} = 1$, while $\sum_{i=1}^{m} x_{ij} \leq 1$ continues to hold for all other j. This phase is iterated until either the solution becomes feasible or the required conditions fail.

The following procedure, ADJUST, is an efficient implementation of the multiplier adjustment method. After the initial solution has been determined, a heuristic phase attempts to satisfy violated constraints (7.3) through pairs (i, j) such that $p_{ij} - \lambda_j = 0$. The adjustment phase then considers items j^* violating (7.3) and computes, for $i \in M$, the least decrease Δ_{ij^*} required in λ_{j^*} for item j^* to be included in the optimal solution to KP_i^{λ}. If an item j^* is found for which

(a) $\min_2 \{\Delta_{1,j^*}, \ldots, \Delta_{mj^*}\} > 0$;

(b) decreasing λ_{j^*} by $\min_2 \{\Delta_{1,j^*}, \ldots, \Delta_{mj^*}\}$ the new Lagrangian solution satisfies $\sum_{i=1}^{m} x_{ij} \leq 1$ for all $j \in N$,

then such updating is performed (decreasing the current upper bound value by $\min\{\Delta_{1,j^*}, \ldots, \Delta_{mj^*}\}$) and a new heuristic phase is attempted. If no such j^* exists, the process terminates.

The output variables define the upper bound value

$$U_3 = \sum_{i=1}^{m} z_i + \sum_{j=1}^{n} \lambda_j ; \tag{7.19}$$

if *opt* = "yes", this value is optimal and (x_{ij}) gives the corresponding solution.

```
procedure ADJUST :
input: n, m, (p_ij), (w_ij), (c_i);
output: (z_i), (λ_j), (x_ij), U_3, opt;
begin
    comment: initialization;
    N := {1, ..., n};
    M := {1, ..., m};
    for i := 1 to m do for j := 1 to n do x_ij := 0;
    for j := 1 to n do λ_j := max_2{ p_ij : i ∈ M, w_ij ≤ c_i };
    U_3 := Σ_{j∈N} λ_j;
    for i := 1 to m do
        begin
            N_i := {j ∈ N : p_ij − λ_j > 0};
            set x_ij (j ∈ N_i) to the solution to
                max  z_i = Σ_{j∈N_i} (p_ij − λ_j) x_ij
                subject to Σ_{j∈N_i} w_ij x_ij ≤ c_i ,
                           x_ij = 0 or 1, j ∈ N_i;
```

```
            U_3 := U_3 + z_i
        end;
    opt := "no";
    if Σ_{i∈M} x_ij = 1 for all j ∈ N then opt := "yes"
    else
        repeat
            comment: heuristic phase;
            IJ := {(i, j) : Σ_{k∈M} x_kj = 0, p_ij − λ_j = 0};
            for each (i, j) ∈ IJ, in order of decreasing p_ij, do
                if Σ_{k∈M} x_kj = 0 and w_ij + Σ_{l∈N} w_il x_il ≤ c_i then x_ij := 1;
            comment: adjustment ;
            if Σ_{i∈M} x_ij = 1 for all j ∈ N then opt := "yes"
            else
                begin
                    J := {j ∈ N : Σ_{k∈M} x_kj = 0};
                    found := "no";
                    repeat
                        let j* be any index in J;
                        J := J\{j*} ;
                        M_j* := {i ∈ M : w_ij* ≤ c_i};
                        for each i ∈ M_j* do
                            begin
                                N_i := {j ∈ N\{j*} : p_ij − λ_j > 0, w_ij ≤ c_i − w_ij*};
                                determine the solution to
                                (KP_i) max z̃_i = Σ_{j∈N_i} (p_ij − λ_j) y_j
                                        subject to Σ_{j∈N_i} w_ij y_j ≤ c_i − w_ij*,
                                                   y_j = 0 or 1, j ∈ N_i;
                                Δ_ij* := z_i − (z̃_i + (p_ij* − λ_j*))
                            end;
                        if min_2{Δ_ij* : i ∈ M_j*} > 0 then
                            begin
                                i* := arg min {Δ_ij* : i ∈ M_j*};
                                let (y_j), j ∈ N_i*, be the solution found for KP_i*;
                                for each j ∈ N\N_i* do y_j := 0;
                                y_j* := 1;
                                if y_j + Σ_{i∈M\{i*}} x_ij ≤ 1 for all j ∈ N then
                                    begin
                                        found := "yes";
                                        λ_j* := λ_j* − min_2{Δ_ij* : i ∈ M_j*};
                                        replace row i* of x with (y_j);
                                        z_i* := z̃_i* + (p_i*j* − λ_j*);
                                        U_3 := U_3 − Δ_i*j*
                                    end
                            end
                    until J = ∅ or found = "yes"
                end
        until opt = "yes" or found = "no"
end.
```

Example 7.2 (continued)

The initial solution is obtained by setting

$(\lambda_j) = (5, 3, 3, 4, 1)$:

$z_1 = 10, (x_{1,j}) = (0, 0, 0, 1, 1);$

$z_2 = 5, (x_{2,j}) = (0, 0, 1, 0, 0);$

$U_3 = 16 + (10 + 5) = 31 = U_1.$

The heuristic phase has no effect, hence $J = \{1, 2\}$. For $j^* = 1$ we obtain

$M_1 = \{1, 2\};$

$N_1 = \{5\}, \tilde{z}_1 = 6, y_5 = 1, \Delta_{1,1} = 2;$

$N_2 = \emptyset, \tilde{z}_2 = 0, \Delta_{2,1} = 5,$

hence $i^* = 1$. Replacing $(x_{1,j})$ with (1, 0, 0, 0, 1), condition $\sum_{i \in M} x_{ij} \leq 1$ continues to hold for all $j \in N$, so we have

$(\lambda_j) = (0, 3, 3, 4, 1);$

$z_1 = 13 , (x_{1,j}) = (1, 0, 0, 0, 1);$

$U_3 = 29.$

The heuristic phase sets $x_{1,2} = 1$, hence $J = \{4\}$. For $j^* = 4$ we have

$M_4 = \{1, 2\};$

$N_1 = \{5\}, \tilde{z}_1 = 6, y_5 = 1, \Delta_{1,4} = 3;$

$N_2 = \{1\}, \tilde{z}_2 = 5, y_1 = 1, \Delta_{2,4} = 0,$

so the execution terminates with $U_3 = 29$. For this instance we have $U_3 < U_1 (= 31)$, but $U_3 > \overline{U}_0 (= 26) > U_2 (= 25)$. On the other hand, applying procedure ADJUST to the instance of Example 7.1, we initially have $U_3 = 45$, then the first adjustment improves it to 43 and the second to 42 (with two further adjustments producing no improvement). Hence $U_3 = 42 < U_2 (= 49) < \overline{U}_0 (= 54)$. □

Examples 7.1 and 7.2 prove that no dominance exists between the Fisher–Jaikumar–Van Wassenhove bound (U_3) and the Martello–Toth bounds ($\overline{U}_0$ and

U_2), nor between the Ross–Soland (U_0 and U_1) and the Martello–Toth bounds. As already shown, the only dominances among these bounds are $U_3 \le U_1 \le U_0$ and $U_2 \le \overline{U}_0$.

7.2.4 The variable splitting method

Jörnsten and Näsberg (1986) have introduced a new way of relaxing GAP in a Lagrangian fashion. (A general discussion on this kind of relaxation can be found in Guignard and Kim (1987).) By introducing extra binary variables y_{ij} ($i \in M$, $j \in N$) and two positive parameters α and β, the problem is formulated, through *variable splitting*, as

$$\text{maximize} \quad \alpha \sum_{i=1}^{m} \sum_{j=1}^{n} p_{ij} x_{ij} + \beta \sum_{i=1}^{m} \sum_{j=1}^{n} p_{ij} y_{ij} \tag{7.20}$$

$$\text{subject to} \quad \sum_{j=1}^{n} w_{ij} x_{ij} \le c_i, \quad i \in M, \tag{7.21}$$

$$\sum_{i=1}^{m} y_{ij} = 1, \quad j \in N, \tag{7.22}$$

$$x_{ij} = y_{ij}, \quad i \in M, j \in N, \tag{7.23}$$

$$x_{ij} = 0 \text{ or } 1, \quad i \in M, j \in N, \tag{7.24}$$

$$y_{ij} = 0 \text{ or } 1, \quad i \in M, j \in N. \tag{7.25}$$

We denote problem (7.20)–(7.25) by XYGAP. It is immediate that XYGAP is equivalent to GAP in the sense that the corresponding optimal solution values, $z(XYGAP)$ and $z(GAP)$, satisfy

$$z(XYGAP) = (\alpha + \beta)\, z(GAP). \tag{7.26}$$

The new formulation appears less natural than the original one, but it allows a relaxation of constraints (7.23) through Lagrangian multipliers (μ_{ij}). The resulting problem, $L(XYGAP, \mu)$,

$$\text{maximize} \quad \alpha \sum_{i=1}^{m} \sum_{j=1}^{n} p_{ij} x_{ij} + \beta \sum_{i=1}^{m} \sum_{j=1}^{n} p_{ij} y_{ij} + \sum_{i=1}^{m} \sum_{j=1}^{n} \mu_{ij} (x_{ij} - y_{ij}) \tag{7.27}$$

subject to (7.21), (7.22), (7.24), (7.25),

keeps both sets of GAP constraints, and immediately separates into two problems, one, $XGAP(\mu)$, in the x variables and one, $YGAP(\mu)$, in the y variables. The former,

$$\text{maximize} \quad z(XGAP(\mu)) = \sum_{i=1}^{m} \sum_{j=1}^{n} (\alpha p_{ij} + \mu_{ij}) x_{ij}$$

$$\text{subject to} \quad \sum_{j=1}^{n} w_{ij} x_{ij} \leq c_i, \qquad i \in M,$$

$$x_{ij} = 0 \text{ or } 1, \qquad i \in M, j \in N,$$

has the same structure as $L(GAP, \lambda)$ (Section 7.2.1), hence separates into m 0-1 single knapsack problems (KP_i^{μ}, $i = 1, \dots, m$) of the form

$$\text{maximize} \quad \hat{z}_i = \sum_{j=1}^{n} (\alpha p_{ij} + \mu_{ij}) x_{ij}$$

$$\text{subject to} \quad \sum_{j=1}^{n} w_{ij} x_{ij} \leq c_i,$$

$$x_{ij} = 0 \text{ or } 1, \quad j \in N;$$

the latter

$$\text{maximize} \quad z(YGAP(\mu)) = \sum_{i=1}^{m} \sum_{j=1}^{n} (\beta p_{ij} - \mu_{ij}) y_{ij}$$

$$\text{subject to} \quad \sum_{i=1}^{m} y_{ij} = 1, \qquad j \in N,$$

$$y_{ij} = 0 \text{ or } 1, \qquad i \in M, j \in N,$$

has the same structure as the initial Ross–Soland relaxation (Section 7.2.1), hence its optimal solution is

$$y_{ij} = \begin{cases} 1 & \text{if } i = i(j); \\ 0 & \text{otherwise,} \end{cases} \qquad \text{for } j \in N,$$

where

$$i(j) = \arg\max \{\beta p_{ij} - \mu_{ij} : i \in M, w_{ij} \leq c_i\}.$$

By solving problems KP_i^{μ} ($i \in M$), we obtain the solution to $L(XYGAP, \mu)$, of value

$$z(L(XYGAP,\mu)) = \sum_{i=1}^{m} \hat{z}_i + \sum_{j=1}^{n} (\beta p_{i(j),j} - \mu_{i(j),j}), \tag{7.28}$$

hence the upper bound

$$U_4 = \lfloor z(L(XYGAP,\mu))/(\alpha+\beta) \rfloor. \tag{7.29}$$

Jörnsten and Näsberg (1986) have proved that, for $\alpha + \beta = 1$ and for the *optimal* Lagrangian multipliers λ^*, μ^*,

$$z(L(XYGAP,\mu^*)) \leq z(L(GAP,\lambda^*)).$$

However, there is no analytical way to determine μ^*, and the multiplier adjustment method of Section 7.2.3 does not appear adaptable to XYGAP. Jörnsten and Näsberg have proposed using a subgradient optimization algorithm to determine a "good" μ. At each iteration, the current μ is updated by setting $\mu_{ij} = \mu_{ij} + t(y_{ij} - x_{ij})$ $(i \in M, j \in N)$, where t is an appropriate positive step.

Example 7.2 (continued)

Using $\alpha = \beta = \frac{1}{2}$ and starting with $\mu_{ij} = 0$ for all $i \in M$, $j \in N$, we obtain

$$(x_{ij}) = \begin{pmatrix} 1 & 1 & 0 & 0 & 1 \\ 1 & 0 & 0 & 1 & 0 \end{pmatrix},$$

i.e., the same solution found for $\overline{U}_0$ (Section 7.2.2), and

$$(y_{ij}) = \begin{pmatrix} 1 & 1 & 0 & 1 & 1 \\ 0 & 0 & 1 & 0 & 0 \end{pmatrix},$$

i.e., the same solution found for U_0. The initial upper bound value is thus

$$U_4 = \lfloor 13 + 16.5 \rfloor = 29 \ (= U_3).$$

Assuming that the initial step is $t = 1$, we then have

$$(\mu_{ij}) = \begin{pmatrix} 0 & 0 & 0 & 1 & 0 \\ -1 & 0 & 1 & -1 & 0 \end{pmatrix};$$

$$(\alpha p_{ij} + \mu_{ij}) = \begin{pmatrix} \frac{7}{2} & \frac{3}{2} & \frac{3}{2} & 5 & \frac{7}{2} \\ \frac{3}{2} & \frac{3}{2} & 5 & 1 & \frac{1}{2} \end{pmatrix};$$

$$(x_{ij}) = \begin{pmatrix} 1 & 1 & 0 & 0 & 1 \\ 0 & 0 & 1 & 0 & 0 \end{pmatrix};$$

$$(\beta p_{ij} - \mu_{ij}) = \begin{pmatrix} \frac{7}{2} & \frac{3}{2} & \frac{3}{2} & 3 & \frac{7}{2} \\ \frac{7}{2} & \frac{3}{2} & 3 & 3 & \frac{1}{2} \end{pmatrix};$$

$$(y_{ij}) = \begin{pmatrix} 1 & 1 & 0 & 1 & 1 \\ 0 & 0 & 1 & 0 & 0 \end{pmatrix},$$

and the upper bound becomes

$$U_4 = \lfloor 13.5 + 14.5 \rfloor = 28.$$

Further improvements could be obtained by iterating the procedure. □

7.3 EXACT ALGORITHMS

The most commonly used methods in the literature for the exact solution of GAP are depth-first branch-and-bound algorithms.

In the Ross and Soland (1975) scheme, upper bound U_1 (see Section 7.2.1) is computed at each node of the branch-decision tree. The branching variable is selected through the information determined for computing U_1. In fact, the variable chosen to separate, $x_{i^*j^*}$, is the one, among those with $y_{ij} = 0$ ($i \in M'$, $j \in N'$) in the optimal solution to problems KP_i^1 ($i \in M'$), for which the quantity

$$\frac{q_j}{w_{ij} / \left(c_i - \sum_{k=1}^{n} w_{ik} x_{ik}\right)}$$

is a maximum. This variable represents an item j^* which is "well fit" into knapsack i^*, considering both the penalty for re-assigning the item and the residual capacity of the knapsack. Two branches are then generated by imposing $x_{i^*j^*} = 1$ and $x_{i^*j^*} = 0$.

In the Martello and Toth (1981c) scheme, upper bound min (U_1, U_2) (see Sections 7.2.1, 7.2.2) is computed at each node of the branch-decision tree. In addition, at the root node, a tighter upper bound on the global solution is determined by computing min (U_3, U_2) (see Section 7.2.3). The information computed for U_2 determines the branching as follows. The separation is performed on item

$$j^* = \arg\max \{l_j : j \in N^0 \cup N^>\},$$

i.e. on the item whose re-assignment is likely to produce the maximum decrease of the objective function. If $j^* \in N^0$, m nodes are generated by assigning j^* to each knapsack in turn (as shown in Figure 7.1(a)); if $j^* \in N^>$, with $M^>(j^*) = \{i_1, i_2, \ldots, i_{\overline{m}}\}$, $\overline{m} - 1$ nodes are generated by assigning j^* to knapsacks $i_1, \ldots, i_{\overline{m}-1}$ in turn, and another node by excluding j^* from knapsacks $i_1, \ldots, i_{\overline{m}-1}$ (as shown in Figure 7.1(b)). With this branching strategy, m single knapsack problems KP_i^2 must be solved to compute the upper bound associated with the root node, but only one new KP_i^2 for each other node of the tree. In fact if $j^* \in N^0$, imposing $x_{kj^*} = 1$ requires only the solution of problem KP_k^2, the solutions to problems KP_i^2 $(i \neq k)$ being unchanged with respect to the generating node; if $j^* \in N^>$, the strategy is the same as that used in the Martello and Toth (1980a) algorithm for the 0-1 multiple knapsack problem (see Section 6.4.1), for which we have shown that the solution of $\overline{m}$ problems KP_i^2 produces the upper bounds corresponding to the $\overline{m}$ generated nodes.

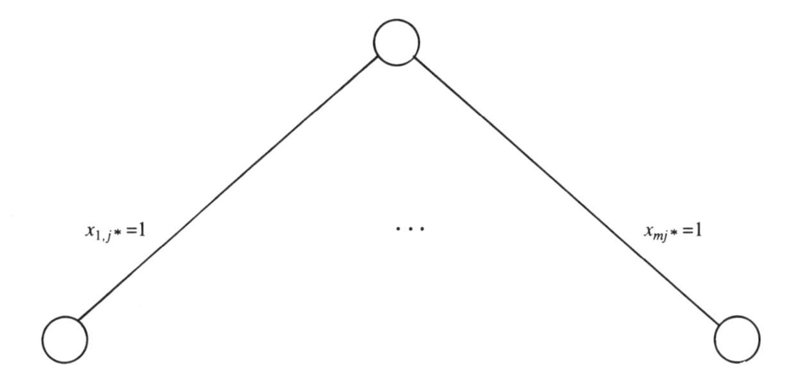

Figure 7.1(a) Branching strategy when $j^* \in N^0$

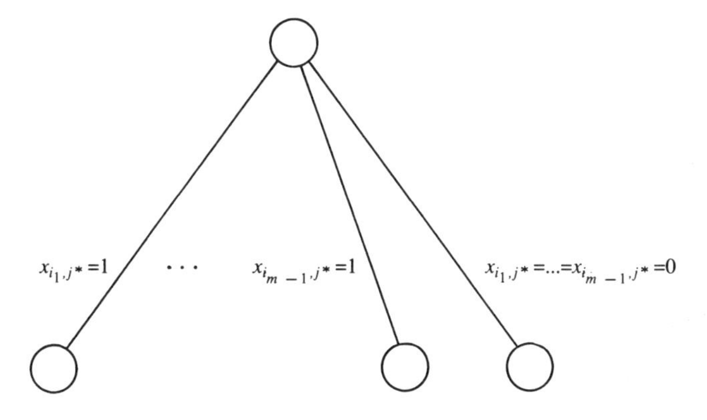

Figure 7.1(b) Branching strategy when $j^* \in N^{>}$

The execution of the above scheme is preceded by a preprocessing which: (a) determines an approximate solution through a procedure, MTHG, described in the next section; (b) reduces the size of the instance, through two procedures, MTRG1 and MTRG2, described in Section 7.5. (Example 7.3 of Section 7.5 illustrates the branching scheme.) At each decision node, a partial reduction is performed, by searching for unassigned items which can currently be assigned to only one knapsack. The Fortran implementation of the resulting algorithm (MTG) is included in the present volume.

In the Fisher, Jakumar and Van Wassenhove (1986) scheme, upper bound U_3 (Section 7.2.3) is computed at each node of the branch-decision tree. The branching variable is an $x_{i^*j^*}$ corresponding to a $w_{i^*j^*}$ which is maximum over all variables that have not been fixed to 0 or 1 at previous branches. Two nodes are then generated by fixing $x_{i^*j^*} = 1$ and $x_{i^*j^*} = 0$.

No scheme has been proposed by Jörnsten and Näsberg (1986).

7.4 APPROXIMATE ALGORITHMS

As seen in Section 7.1, determining whether an instance of GAP (or MINGAP) has a feasible solution is an NP-complete problem. It follows that, unless $\mathcal{P} = \mathcal{NP}$, these problems admit no polynomial-time approximate algorithm with fixed worst-case performance ratio, hence also no polynomial-time approximation scheme.

The following polynomial-time algorithm (Martello and Toth, 1981c) provides approximate solutions to GAP. Let f_{ij} be a measure of the "desirability" of assigning item j to knapsack i. We iteratively consider all the unassigned items, and determine the item j^* having the maximum difference between the largest and the second largest f_{ij} $(i \in M)$; j^* is then assigned to the knapsack for which f_{ij^*} is a maximum. In the second part of the algorithm the current solution is improved through local exchanges. On output, if *feas* = "no", no feasible solution has been determined; otherwise the solution found, of value z^h, is stored in y_j = (knapsack to which item j is assigned), $j = 1, \dots, n$.

procedure MTHG:
input: $n, m, (p_{ij}), (w_{ij}), (c_i), (f_{ij})$;
output: $z^h, (y_j)$, *feas*;
begin
  $M := \{1, \dots, m\}$;
  $U := \{1, \dots, n\}$;
  comment: initial solution;
  feas := "yes"
  for $i := 1$ **to** m **do** $\overline{c}_i := c_i$;
  $z^h := 0$;
  while $U \neq \emptyset$ and *feas* = "yes" **do**
    begin
      $d^* := -\infty$;
      for each $j \in U$ **do**
        begin
          $F_j := \{i \in M : w_{ij} \leq \overline{c}_i\}$;
          if $F_j = \emptyset$ **then** *feas* := "no"
          else

```
                  begin
                      i' := arg max { f_ij : i ∈ F_j };
                      if F_j\{i'} = Ø then d := +∞
                      else d := f_i'j − max2{ f_ij : i ∈ F_j };
                      if d > d* then
                          begin
                              d* := d;
                              i* := i';
                              j* := j
                          end
                  end
          end;
      if feas = "yes" then
          begin
              y_j* := i*;
              z^h := z^h + p_i*j*;
              c̄_i* := c̄_i* − w_i*j*;
              U := U\{j*}
          end
  end;
comment: improvement;
if feas = "yes" then
  for j := 1 to n do
      begin
          i' := y_j;
          A := { p_ij : i ∈ M\{i'}, w_ij ≤ c̄_i };
          if A ≠ Ø then
              begin
                  let p_i''j = max A;
                  if p_i''j > p_i'j then
                      begin
                          y_j := i'';
                          z^h := z^h − p_i'j + p_i''j;
                          c̄_i' := c̄_i' + w_i'j;
                          c̄_i'' := c̄_i'' − w_i''j
                      end
              end
      end
end.
```

Procedure MTHG can be implemented efficiently by initially sorting in decreasing order, for each item j, the values f_{ij} $(i \in M)$ such that $w_{ij} \leq \overline{c}_i$ $(= c_i)$. This requires $O(nm \log m)$ time, and makes immediately available, at each iteration in the inner loop, the pointers to the maximum and the second maximum element of $\{ f_{ij} : i \in F_j \}$. Hence the main while loop performs the $O(n)$ assignments within a total of $O(n^2)$ time. Whenever an item is assigned, the decrease in $\overline{c}_{i^*}$ can make it necessary to update the pointers. Since, however, the above maxima can only decrease during execution, a total of $O(n^2)$ operations is required by the algorithm for these checks and updatings. By finally observing that the exchange phase clearly takes $O(nm)$ time, we conclude that the overall time complexity of MTHG is $O(nm\log m + n^2)$.

Computational experiments have shown that good results can be obtained using the following choices for f_{ij}:

(a) $f_{ij} = p_{ij}$ (with this choice the improvement phase can be skipped);
(b) $f_{ij} = p_{ij}/w_{ij}$;
(c) $f_{ij} = -w_{ij}$;
(d) $f_{ij} = -w_{ij}/c_i$.

Example 7.3

Consider the instance of GAP defined by

$$n = 8;$$

$$m = 3;$$

$$(p_{ij}) = \begin{pmatrix} 27 & 12 & 12 & 16 & 24 & 31 & 41 & 13 \\ 14 & 5 & 37 & 9 & 36 & 25 & 1 & 34 \\ 34 & 34 & 20 & 9 & 19 & 19 & 3 & 34 \end{pmatrix};$$

$$(w_{ij}) = \begin{pmatrix} 21 & 13 & 9 & 5 & 7 & 15 & 5 & 24 \\ 20 & 8 & 18 & 25 & 6 & 6 & 9 & 6 \\ 16 & 16 & 18 & 24 & 11 & 11 & 16 & 18 \end{pmatrix}; \ (c_i) = \begin{pmatrix} 26 \\ 25 \\ 34 \end{pmatrix}.$$

Let us consider execution of MTHG with $f_{ij} = -w_{ij}$. The first phase of the algorithm gives

$$j^* = 4 : d^* = 19, \quad y_4 = 1, \overline{c}_1 = 21;$$

$$j^* = 8 : d^* = 12, \quad y_8 = 2, \overline{c}_2 = 19;$$

$$j^* = 3 : d^* = 9, \quad y_3 = 1, \overline{c}_1 = 12;$$

$$j^* = 1 : d^* = +\infty, y_1 = 3, \overline{c}_3 = 18;$$

$$j^* = 2 : d^* = 8, \quad y_2 = 2, \overline{c}_2 = 11;$$

$$j^* = 6 : d^* = 5, \quad y_6 = 2, \overline{c}_2 = 5;$$

$$j^* = 7 : d^* = 11, \quad y_7 = 1, \overline{c}_1 = 7;$$

$$j^* = 5 : d^* = 4, \quad y_5 = 1, \overline{c}_1 = 0;$$

hence

$$z^h = 191, (y_j) = (3, 2, 1, 1, 1, 2, 1, 2), (\overline{c}_i) = (0, 5, 18).$$

The second phase performs the exchanges

$$j = 2 : y_2 = 3, \overline{c}_2 = 13, \overline{c}_3 = 2;$$
$$j = 5 : y_5 = 2, \overline{c}_1 = 7, \overline{c}_2 = 7;$$

so the solution found is

$$z^h = 232, (y_j) = (3, 3, 1, 1, 2, 2, 1, 2). \square$$

A Fortran implementation of MTHG, which determines the best solution obtainable with choices (a)–(d) for f_{ij}, is included in the present volume. A more complex approximate algorithm, involving a modified subgradient optimization approach and branch-and-bound, can be found in Klastorin (1979).

Mazzola (1989) has derived from MTHG an approximate algorithm for the generalization of GAP arising when the capacity constraints (7.2) are non-linear.

7.5 REDUCTION ALGORITHMS

The following algorithms (Martello and Toth, 1981c) can be used to reduce the size of an instance of GAP. Let (y_j) define a feasible solution (determined, for example, by procedure MTHG of the previous section) of value $z^h = \sum_{j=1}^{n} p_{y_j, j}$.

The first reduction algorithm receives in input the upper bound value U_0 of Section 7.2.1 and the corresponding values $i(j) = \arg\max \{p_{ij} : i \in M, w_{ij} \leq c_i\}$ $(j \in N)$. The algorithm fixes to 0 those variables x_{ij} which, if set to 1, would decrease the bound to a value not greater than z^h. (We obviously assume $z^h < U_0$.) If, for some j, all x_{ij} but one, say x_{i^*j}, are fixed to 0, then x_{i^*j} is fixed to 1. We assume that, on input, all entries of (x_{ij}) are preset to a dummy value other than 0 or 1. On output, k_j^0 $(j \in N)$ has the value $|\{x_{ij} : i \in M, x_{ij} = 0\}|$, and $\overline{c}_i$ gives the residual capacity of knapsack i $(i \in M)$; these values are used by the second reduction algorithm. We also assume that, initially, $\overline{c}_i = c_i$ for all $i \in M$. If, for some j, all x_{ij} are fixed to 0, the feasible solution (y_j) is optimal, hence the output variable *opt* takes the value "yes".

procedure MTRG1:
input: $n, m, (p_{ij}), (w_{ij}), (\overline{c}_i), z^h, U_0, (i(j)), (x_{ij})$;
output: $(x_{ij}), (k_j^0), (\overline{c}_i)$, *opt* ;
begin
    opt := "no";
    j := 0;
    while $j < n$ and *opt* = "no" **do**
        begin
            $j := j + 1$;
            $k_j^0 := 0$;
            for i := 1 **to** m **do**
                if $z^h \geq U_0 - p_{i(j),j} + p_{ij}$ or $w_{ij} > \overline{c}_i$ **then**

```
                    begin
                        x_ij := 0;
                        k_j^0 := k_j^0 + 1
                    end
                else i* := i;
            if k_j^0 = m − 1 then
                begin
                    x_i*j := 1;
                    c̄_i* := c̄_i* − w_i*j
                end
            else if k_j^0 = m then opt := "yes"
        end
end.
```

The time complexity of MTRG1 is clearly $O(nm)$. When the execution fixes some variable to 1, hence decreasing some capacity $\overline{c}_i$, further reductions can be obtained by reapplying the procedure. Since n variables at most can be fixed to 1, the resulting time complexity is $O(n^2m)$.

The second reduction algorithm receives in input (x_{ij}), (k_j^0), $(\overline{c}_i)$, the upper bound value $\overline{U}_0$ of Section 7.2.2 and, for each problem KP_i^2 $(i \in M)$, the corresponding solution $(\hat{x}_{i,1}, \ldots, \hat{x}_{in})$ and optimal value z_i. Computation of the upper bounds of Section 7.2.2,

$$u_{ij}^0 = \text{ current upper bound on the solution value of } KP_i^2 \text{ if } x_{ij} = 0;$$

$$u_{ij}^1 = \text{ current upper bound on the solution value of } KP_i^2 \text{ if } x_{ij} = 1,$$

is then used to fix to $\hat{x}_{ij}$ variables x_{ij} which, if set to $1 - \hat{x}_{ij}$, would give an upper bound not greater than z^h. We assume that MTRG1 is first iteratively run, then $\overline{U}_0$ and the solutions to problems KP_i^2 are determined using the reductions obtained. Consequently, the new algorithm cannot take decisions contradicting those of MTRG1. It can, however, fix to 1 more than one variable in a column, or to 0 all the variables in a column. Such situations imply that the current approximate solution is optimal, hence the output variable *opt* takes the value "yes".

```
procedure MTRG2:
input: n, m, (p_ij), (w_ij), (c̄_i), z^h, Ū_0, (z_i), (x̂_ij), (x_ij), (k_j^0);
output: (x_ij), opt;
begin
    opt := "no";
    j := 1;
    repeat
        if k_j^0 < m − 1 then
            begin
                k1 := 0;
                for i := 1 to m do
                    if x_ij ≠ 0 then
```

```
                    if w_ij > c̄_i then
                        begin
                            x_ij := 0;
                            k_j^0 := k_j^0 + 1
                        end
                    else
                        if x̂_ij = 0  and  z^h ≥ Ū_0 − z_i + u_ij^1 then
                            begin
                                x_ij := 0;
                                k_j^0 := k_j^0 + 1
                            end
                        else
                            begin
                                if k1 = 0 then i* := i;
                                if x̂_ij = 1  and  z^h ≥ Ū_0 − z_i + u_ij^0 then
                                    if k1 = 0 then k1 := 1
                                    else opt := "yes"
                            end;
            if opt = "no" then
                if k_j^0 = m − 1  or  k1 = 1 then
                    begin
                        for i := 1 to m do x_ij := 0;
                        x_i*j := 1;
                        c̄_i* := c̄_i* − w_i*j;
                        k_j^0 := m − 1
                    end
                else
                    if k_j^0 = m then opt = "yes"
        end;
        j := j + 1
    until j > n  or  opt = "yes"
end.
```

If u_{ij}^0 and u_{ij}^1 are computed through any of the $O(n)$ methods of Sections 2.2 and 2.3, the time complexity of MTRG2 is $O(mn^2)$. In this case too, when a variable has been fixed to 1, a new execution can produce further reductions.

Example 7.3 (continued)

Using the solution value $z^h = 232$ found by MTHG and the upper bound value $U_0 = 263$, MTRG1 gives

$j = 7 : x_{2,7} = 0, x_{3,7} = 0$, hence $k_7^0 = 2$, so

$x_{1,7} = 1, \overline{c}_1 = 21;$

$j = 8 : x_{1,8} = 0.$

Solving KP_i^2 $(i = 1, 2, 3)$ for the reduced problem, we get

$$(\hat{x}_{ij}) = \begin{pmatrix} 0 & 0 & 1 & 1 & 1 & 0 & \textcircled{1} & \textcircled{0} \\ 0 & 0 & 0 & 0 & 1 & 1 & \textcircled{0} & 1 \\ 1 & 1 & 0 & 0 & 0 & 0 & \textcircled{0} & 0 \end{pmatrix}, (z_i) = \begin{pmatrix} 93 \\ 95 \\ 68 \end{pmatrix}, \overline{U}_0 = 256,$$

where fixed x_{ij} values are circled. Executing MTRG2 we have

$j = 1 : x_{1,1} = 0, x_{2,1} = 0$, hence $x_{3,1} = 1, \overline{c}_3 = 18$;

$j = 4 : x_{2,4} = 0, x_{3,4} = 0$, hence $x_{1,4} = 1, \overline{c}_1 = 16$.

The execution of the Martello and Toth (1981c) branch-and-bound algorithm (see Section 7.3) follows. Computation of U_2 and U_3 for the root node gives

$N^0 = \emptyset$, $N^> = \{5\}$, $M^>(5) = \{1, 2\}$;

$u^0_{1,5} = 89$, $u^0_{2,5} = 85$; $l_5 = 4$, $U_2 = 252$;

$U_3 = 245$.

The branching scheme is shown in Figure 7.2. Since $j^* = 5$, we generate nodes 1 and 2, and compute the corresponding bounds.

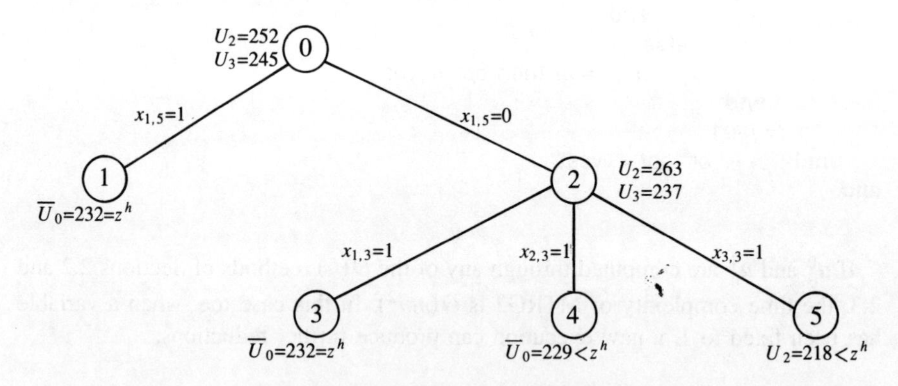

Figure 7.2 Decision-tree for Example 7.3

Node 1 : $(\hat{x}_{2,j}) = (0, 0, 1, 0, 0, 0, 0, 1)$, $z_2 = 71$, $\overline{U}_0 = 232 = z^h$.

Node 2 : $(\hat{x}_{1,j}) = (0, 0, 0, 1, 0, 1, 1, 0)$, $z_1 = 88$, $\overline{U}_0 = 251$;

$N^0 = \{3\}$, $N^> = \{6\}$, $M^>(6) = \{1, 2\}$;

$u^1_{1,3} = 69$, $u^1_{2,3} = 78$, $u^1_{3,3} = 54$, $l_3 = 14$;

$u^0_{1,6} = 75$, $u^0_{2,6} = 96$, $l_6 = 0$;

$$U_2 = 237;$$

$$U_0 = U_1 = 263 \text{ (unchanged)}.$$

The highest penalty is $l_3 = 14$, hence $j^* = 3$ and nodes 3, 4, 5 are generated.

$$\text{Node } 3 : (\hat{x}_{1,j}) = (0, 0, 1, 1, 0, 0, 1, 0), \ z_1 = 69, \ \overline{U}_0 = 232 = z^h.$$

$$\text{Node } 4 : (\hat{x}_{2,j}) = (0, 0, 1, 0, 1, 0, 0, 0), \ z_2 = 73, \ \overline{U}_0 = 229 < z^h.$$

$$\text{Node } 5 : (\hat{x}_{3,j}) = (1, 0, 1, 0, 0, 0, 0, 0), \ z_3 = 54, \ \overline{U}_0 = 237;$$

$$N^0 = \{2\}, \ N^> = \{6\}, \ M^>(6) = \{1, \ 2\};$$

$$u^1_{1,2} = 69, \ u^1_{2,2} = 95, \ u^1_{3,2} = -\infty, \ l_2 = 0;$$

$$u^0_{1,6} = 69, \ u^0_{2,6} = 75, \ l_6 = 19;$$

$$U_2 = 218 < z^h.$$

The approximate solution $(y_1) = (3, 3, 1, 1, 2, 2, 1, 2)$, of value $z^h = 232$, is thus optimal. □

7.6 COMPUTATIONAL EXPERIMENTS

Tables 7.1 to 7.4 compare the exact algorithms of Section 7.3 on four classes of randomly generated problems. For the sake of uniformity with the literature (Ross and Soland (1975), Martello and Toth (1981c), Fisher, Jaikumar and Van Wassenhove (1986)), all generated instances are *minimization* problems of the form (7.5), (7.2), (7.3), (7.4). All the algorithms we consider except the Ross and Soland (1975) one, solve maximization problems, so the generated instances are transformed through (7.7), using for t the value

$$t = \max_{i \in M, \ j \in N} \{c_{ij}\} + 1.$$

The classes are

(a) w_{ij} uniformly random in [5, 25],
c_{ij} uniformly random in [1, 40],
$c_i = 9(n/m) + 0.4 \max_{i \in M} \{\sum_{j \in N_i} w_{ij}\}$ for $i = 1, \dots, m$
(where N_i is defined as in Section 7.2.1);

(b) w_{ij} and c_{ij} as for class (a),
$c_i = 0.7(9(n/m) + 0.4 \max_{i \in M} \{\sum_{j \in N_i} w_{ij}\})$ for $i = 1, \dots, m$;

(c) w_{ij} and c_{ij} as for class (a),
$c_i = 0.8 \sum_{j=1}^{n} w_{ij}/m$ for $i = 1, \dots, m$;

Table 7.1 Data set (a). HP 9000/840 in seconds. Average times/Average numbers of nodes over 10 problems

		RS		MTG		FJV		MTGFJV		MTGJN	
m	n	Time	Nodes	Time	Nodes	Time	Nodes	Time	Nodes	Time	Nodes
	10	0.003	3	0.014	1	0.010	2	0.022	1	0.054	1
2	20	0.020	10	0.038	2	1.792	113	0.042	2	0.152	2
	30	0.018	5	0.060	2	10.089(9)	54	0.086	2	0.283	2
	10	0.006	2	0.016	1	0.005	1	0.018	1	0.019	1
3	20	0.035	18	0.124	16	5.927	293	0.100	5	1.004	15
	30	0.045	15	0.075	1	0.124	6	0.071	1	0.069	1
	10	0.017	10	0.067	7	0.056	3	0.074	3	0.387	7
5	20	0.057	19	0.258	41	6.799	320	0.190	7	1.837	41
	30	0.044	10	0.224	12	0.117	6	0.150	3	1.108	12

Table 7.2 Data set (b). HP 9000/840 in seconds. Average times/Average numbers of nodes over 10 problems

		RS		MTG		FJV		MTGFJV		MTGJN	
m	n	Time	Nodes	Time	Nodes	Time	Nodes	Time	Nodes	Time	Nodes
2	10	0.089	93	0.123	32	0.416	15	0.312	14	0.793	26
	20	13.906	9907	14.827	4012	79.951(5)	714	49.628(8)	630	69.573(6)	2231
	30	time limit		time limit		time limit		time limit		time limit	
3	10	0.157	165	0.228	54	0.925	69	0.539	31	1.312	41
	20	48.804(6)	32264	32.487(7)	5982	time limit		42.530(7)	489	39.231(7)	853
	30	93.695(1)	56292	63.842(6)	6824	time limit		90.755(2)	1003	82.991(3)	1108
5	10	0.314	259	0.454	94	2.094	110	0.565	33	2.739	74
	20	17.036(9)	9194	19.536	2311	92.831(3)	2043	24.755	551	75.723(3)	944
	30	74.580(3)	33854	68.075(4)	5950	81.932(2)	438	80.615(2)	925	80.587(2)	693

Table 7.3 Data set (c). HP 9000/840 in seconds. Average times/Average numbers of nodes over 10 problems

		RS		MTG		FJV		MTGFJV		MTGJN	
m	n	Time	Nodes	Time	Nodes	Time	Nodes	Time	Nodes	Time	Nodes
	10	0.110	121	0.176	50	0.575	22	0.507	25	1.156	44
2	20	17.608	13637	18.717	4788	96.525(4)	1242	48.471(7)	760	44.267(7)	1614
	30	98.142(1)	68620	83.363(2)	13362	time limit		96.024(1)	853	86.580(2)	1985
	10	0.245	240	0.243	43	1.648	92	0.576	28	0.959	19
3	20	47.476(6)	32727	18.215(9)	4106	time limit		33.769(8)	534	31.446(8)	826
	30	92.670(1)	52612	47.820(7)	4924	time limit		91.459(2)	951	67.374(5)	830
	10	0.474	393	0.794	226	4.028	195	1.708	70	6.881	216
5	20	61.614(6)	31328	15.698	1332	time limit		47.882	514	59.436(7)	524
	30	time limit		time limit		time limit		time limit		time limit	

Table 7.4 Data set (d). HP 9000/840 in seconds. Average times/Average numbers of nodes over 10 problems

		RS		MTG		FJV		MTGFJV		MTGJN	
m	n	Time	Nodes	Time	Nodes	Time	Nodes	Time	Nodes	Time	Nodes
2	10	0.150	162	0.168	30	0.254	30	0.364	25	0.760	14
	20	50.575	42321	14.344	2275	86.809(5)	3362	62.123(7)	1405	21.322	290
	30	time limit		79.582(3)	8230	time limit		97.681(1)	1475	95.702(1)	1076
3	10	0.541	575	0.350	48	0.966	80	0.870	41	2.244	38
	20	time limit		16.890	1587	time limit		95.024(2)	876	91.181(3)	801
	30	time limit		97.000(1)	6267	time limit		time limit		time limit	
5	10	0.810	697	0.498	73	1.677	108	1.244	55	3.481	63
	20	time limit		21.203(3)	7722	time limit		time limit		time limit	
	30	time limit		time limit		time limit		time limit		time limit	

(d) w_{ij} uniformly random in [1, 100],
c_{ij} uniformly random in $[w_{ij}, w_{ij} + 20]$,
$c_i = 0.8 \sum_{j=1}^{n} w_{ij} / m$ for $i = 1, \ldots, m$.

Problems of class (a) have been proposed by Ross and Soland (1975) and generally admit many feasible solutions. Problems of classes (b), (c) and (d) have tighter capacity constraints; in addition, in problems of class (d) a correlation between profits and weights (often found in real-world applications) has been introduced.

The entries in the tables give average running times (expressed in seconds) and average numbers of nodes generated in the branch-decision tree. A time limit of 100 seconds was imposed on the running time spent by each algorithm for the solution of a single instance. For data sets for which the time limit occurred, the corresponding entry gives, in brackets, the number of instances solved within 100 seconds (the average values are computed by also considering the interrupted instances). The cases where the time limit occurred for all the instances are denoted as "time limit". The following algorithms have been coded in Fortran IV and run on an HP 9000/840 computer, using option "-o" for the Fortran compiler:

RS = Algorithm of Ross and Soland (1975);

MTG = Algorithm of Martello and Toth (1981c) as described in Section 7.3;

FJV = Algorithm of Fisher, Jaikumar and Van Wassenhove (1986);

MTGFJV = Algorithm MTG with upper bound min (U_2, U_3) (see Sections 7.2.2, 7.2.3) computed at each node of the branch-decision tree;

MTGJN = Algorithm MTG with upper bound min (U_1, U_2, U_4) (see Sections 7.2.1, 7.2.2, 7.2.4) computed at each node of the branch-decision tree.

For all the algorithms, the solution of the 0-1 single knapsack problems was obtained using algorithm MT1 of Section 2.5.2.

For the computation of U_4, needed by MTGJN, the number of iterations in the subgradient optimization procedure was limited to 50—as suggested by the authors (Jörnsten and Näsberg, 1986)—for the root node, and to 10 for the other nodes. The Lagrangian multipliers were initially set to

$$\mu_{ij} = \overline{p} = \frac{1}{m\,n} \sum_{i=1}^{m} \sum_{j=1}^{n} p_{ij}, \quad i \in M, j \in N$$

(as suggested by the authors) for the root node, and to the corresponding values obtained at the end of the previous computation for the other nodes. (Different choices of the number of iterations and of the initial values of the multipliers produced worse computational results.) The step used, at iteration k of the subgradient optimization procedure, to modify the current μ_{ij} values was that proposed by the authors, i.e.

$$t^k = \frac{\overline{p}}{k+1}.$$

The tables show that the fastest algorithms are RS for the "easy" instances of class (a), and MTG for the harder instances (b), (c), (d). Algorithms MTGFJV and MTGJN generate fewer nodes than MTG, but the global running times are larger (the computation of U_3 and U_4 being much heavier than that of U_1 and U_2), mainly for problems of classes (b), (c) and (d).

Algorithm FJV is much worse than the other algorithms for all data sets, contradicting, to a certain extent, the results presented for the same classes of test problems in Fisher, Jaikumar and Van Wassenhove (1986). This could be explained by observing that such results were obtained by comparing executions on different computers and using different random instances. In addition, the current implementation of MTG incorporates, for the root node, the computation of upper bound U_3.

Table 7.5 gives the performance of the Fortran IV implementation of approximate algorithm MTHG (Section 7.4) on large-size instances. The entries give average running times (expressed in seconds) and, in brackets, upper bounds on the average percentage errors. The percentage errors were computed as $100\ (U - z^h)/U$, where $U = \min\ (U_1, U_2, U_3, U_4)$. Only data sets (a), (b) and (c) are considered, since the computation of U for data set (d) required excessive running times. Errors of value 0.000 indicate that all the solutions found were exact. The table shows that the running times are quite small and, with few exceptions, practically independent

Table 7.5 Algorithm MTHG. HP 9000/840 in seconds. Average times (average percentage errors) over 10 problems

m	n	Data set (a)	Data set (b)	Data set (c)
5	50	0.121(0.184)	0.140(5.434)	0.136(6.822)
	100	0.287(0.063)	0.325(4.750)	0.318(5.731)
	200	0.887(0.029)	0.869(4.547)	0.852(6.150)
	500	2.654(0.012)	3.860(5.681)	3.887(6.145)
10	50	0.192(0.016)	0.225(3.425)	0.240(6.243)
	100	0.457(0.019)	0.521(5.160)	0.550(5.908)
	200	1.148(0.004)	1.271(4.799)	1.334(5.190)
	500	3.888(0.006)	5.139(5.704)	5.175(5.553)
20	50	0.393(0.062)	0.399(1.228)	0.438(6.479)
	100	0.743(0.002)	0.866(1.189)	0.888(5.187)
	200	1.693(0.008)	2.011(2.140)	2.035(4.544)
	500	2.967(0.000)	7.442(3.453)	7.351(4.367)
50	50	0.938(0.000)	0.832(0.125)	0.876(2.024)
	100	0.728(0.005)	1.792(0.175)	2.016(4.041)
	200	3.456(0.002)	3.849(0.296)	4.131(3.248)
	500	2.879(0.000)	12.613(0.517)	12.647(3.198)

of the data set. For $n = 500$ and data set (a), the first execution of MTHG (with $f_{ij} = p_{ij}$) almost always produced an optimal solution of value $z^h = U_0$, so the computing times are considerably smaller than for the other data sets. The quality of the solutions found by MTHG is very good for data set (a) and clearly worse for the other data sets, especially for small values of m. However, it is not possible to decide whether these high errors depend only on the approximate solution or also on the upper bound values. Limited experiments indicated that the error computed with respect to the optimal solution value tends to be about half that computed with respect to U.

8

Bin-packing problem

8.1 INTRODUCTION

The *Bin-Packing Problem* (BPP) can be described, using the terminology of knapsack problems, as follows. Given n *items* and n *knapsacks* (or *bins*), with

$$w_j = \textit{weight} \text{ of item } j,$$

$$c = \textit{capacity} \text{ of each bin},$$

assign each item to one bin so that the total weight of the items in each bin does not exceed c and the number of bins used is a minimum. A possible mathematical formulation of the problem is

$$\text{minimize} \quad z = \sum_{i=1}^{n} y_i \tag{8.1}$$

$$\text{subject to} \quad \sum_{j=1}^{n} w_j x_{ij} \leq c y_i, \quad i \in N = \{1, \dots, n\}, \tag{8.2}$$

$$\sum_{i=1}^{n} x_{ij} = 1, \quad j \in N, \tag{8.3}$$

$$y_i = 0 \text{ or } 1, \quad i \in N, \tag{8.4}$$

$$x_{ij} = 0 \text{ or } 1, \quad i \in N, j \in N, \tag{8.5}$$

where

$$y_i = \begin{cases} 1 & \text{if bin } i \text{ is used;} \\ 0 & \text{otherwise,} \end{cases}$$

$$x_{ij} = \begin{cases} 1 & \text{if item } j \text{ is assigned to bin } i; \\ 0 & \text{otherwise.} \end{cases}$$

We will suppose, as is usual, that the weights w_j are positive integers. Hence, without loss of generality, we will also assume that

$$c \quad \text{is a positive integer,} \tag{8.6}$$

$$w_j \leq c \quad \text{for } j \in N. \tag{8.7}$$

If assumption (8.6) is violated, c can be replaced by $\lfloor c \rfloor$. If an item violates assumption (8.7), then the instance is trivially infeasible. There is no easy way, instead, of transforming an instance so as to handle negative weights.

For the sake of simplicity we will also assume that, in any feasible solution, the lowest indexed bins are used, i.e. $y_i \geq y_{i+1}$ for $i = 1, \ldots, n-1$.

Almost the totality of the literature on BPP is concerned with approximate algorithms and their performance. A thorough analysis of such results would require a separate book (the brilliant survey by Coffman, Garey and Johnson (1984), to which the reader is referred, includes a bibliography of more than one hundred references, and new results continue to appear in the literature). In Section 8.2 we briefly summarize the classical results on approximate algorithms. The remainder of the chapter is devoted to lower bounds (Section 8.3), reduction procedures (Section 8.4) and exact algorithms (Section 8.5), on which very little can be found in the literature. Computational experiments are reported in Section 8.6.

8.2 A BRIEF OUTLINE OF APPROXIMATE ALGORITHMS

The simplest approximate approach to the bin packing problem is the *Next-Fit* (NF) algorithm. The first item is assigned to bin 1. Items $2, \ldots, n$ are then considered by increasing indices: each item is assigned to the current bin, if it fits; otherwise, it is assigned to a new bin, which becomes the current one. The time complexity of the algorithm is clearly $O(n)$. It is easy to prove that, for any instance I of BPP, the solution value $NF(I)$ provided by the algorithm satisfies the bound

$$NF(I) \leq 2\, z(I), \tag{8.8}$$

where $z(I)$ denotes the optimal solution value. Furthermore, there exist instances for which the ratio $NF(I)/z(I)$ is arbitrarily close to 2, i.e. the worst-case performance ratio of NF is $r(\text{NF}) = 2$. Note that, for a minimization problem, the *worst-case performance ratio* of an approximate algorithm A is defined as the smallest real number $r(A)$ such that

$$\frac{A(I)}{z(I)} \leq r(A) \text{ for all instances } I,$$

where $A(I)$ denotes the solution value provided by A.

A better algorithm, *First-Fit* (FF), considers the items according to increasing indices and assigns each item to the lowest indexed initialized bin into which it fits; only when the current item cannot fit into any initialized bin, is a new bin introduced. It has been proved in Johnson, Demers, Ullman, Garey and Graham (1974) that

$$FF(I) \leq \frac{17}{10} z(I) + 2 \tag{8.9}$$

for all instances I of BPP, and that there exist instances I, with $z(I)$ arbitrarily large, for which

$$FF(I) > \frac{17}{10} z(I) - 8. \tag{8.10}$$

Because of the constant term in (8.9), as well as in analogous results for other algorithms, the worst-case performance ratio cannot give complete information on the worst-case behaviour. Instead, for the bin packing problem, the *asymptotic worst-case performance ratio* is commonly used. For an approximate algorithm A, this is defined as the minimum real number $r^{\infty}(A)$ such that, for some positive integer k,

$$\frac{A(I)}{z(I)} \leq r^{\infty}(A) \text{ for all instances } I \text{ satisfying } z(I) \geq k;$$

it is then clear, from (8.9)–(8.10), that $r^{\infty}(\text{FF}) = \frac{17}{10}$.

The next algorithm, *Best-Fit* (BF), is obtained from FF by assigning the current item to the feasible bin (if any) having the smallest residual capacity (breaking ties in favour of the lowest indexed bin). Johnson, Demers, Ullman, Garey and Graham (1974) have proved that BF satisfies the same worst-case bounds as FF (see (8.9)–(8.10)), hence $r^{\infty}(\text{BF}) = \frac{17}{10}$.

The time complexity of both FF and BF is $O(n \log n)$. This can be achieved by using a 2–3 tree whose leaves store the current residual capacities of the initialized bins. (A *2–3 tree* is a tree in which: (a) every non-leaf node has 2 or 3 sons; (b) every path from the root to a leaf has the same length l; (c) labels at the nodes allow searching for a given leaf value, updating it, or inserting a new leaf in $O(l)$ time. We refer the reader to Aho, Hopcroft and Ullman (1983) for details on this data structure.) In this way each iteration of FF or BF requires $O(\log n)$ time, since the number of leaves is bounded by n.

Assume now that the items are sorted so that

$$w_1 \geq w_2 \geq \ldots \geq w_n, \tag{8.11}$$

and then NF or FF, or BF is applied. The resulting algorithms, of time complexity $O(n \log n)$, are called *Next-Fit Decreasing* (NFD), *First-Fit Decreasing* (FFD) and *Best-Fit Decreasing* (BFD), respectively. The worst-case analysis of NFD has been done by Baker and Coffman (1981); that of FFD and BFD by Johnson, Demers, Ullman, Garey and Graham (1974), starting from an earlier result of Johnson (1973) who proved that

$$FFD(I) \leq \frac{11}{9} z(I) + 4 \tag{8.12}$$

Table 8.1 Asymptotic worst-case performance ratios of bin-packing algorithms

Algorithm	Time complexity	r^{∞}	$r^{\infty}_{1/2}$	$r^{\infty}_{1/3}$	$r^{\infty}_{1/4}$
NF	$O(n)$	2.000	2.000	1.500	1.333...
FF	$O(n \log n)$	1.700	1.500	1.333...	1.250
BF	$O(n \log n)$	1.700	1.500	1.333...	1.250
NFD	$O(n \log n)$	1.691...	1.424...	1.302...	1.234...
FFD	$O(n \log n)$	1.222...	1.183...	1.183...	1.150
BFD	$O(n \log n)$	1.222...	1.183...	1.183...	1.150

for all instances I. The results are summarized in Table 8.1 (taken from Coffman, Garey and Johnson (1984)), in which the last three columns give, for $\alpha = \frac{1}{2}, \frac{1}{3}, \frac{1}{4}$, the value r^{∞}_{α} of the asymptotic worst-case performance ratio of the algorithms when applied to instances satisfying $\min_{1 \le j \le n} \{w_j\} \le \alpha c$.

8.3 LOWER BOUNDS

Given a lower bounding procedure L for a minimization problem, let $L(I)$ and $z(I)$ denote, respectively, the value produced by L and the optimal solution value for instance I. The *worst-case performance ratio* of L is then defined as the largest real number $\rho(L)$ such that

$$\frac{L(I)}{z(I)} \ge \rho(L) \text{ for all instances } I.$$

8.3.1 Relaxations based lower bounds

For our model of BPP, the continuous relaxation $C(BPP)$ of the problem, given by (8.1)–(8.3) and

$$0 \le y_i \le 1, \qquad i \in N,$$

$$0 \le x_{ij} \le 1, \qquad i \in N, j \in N,$$

can be immediately solved by the values $x_{ii} = 1$, $x_{ij} = 0$ $(j \ne i)$ and $y_i = w_i/c$ for $i \in N$. Hence

$$z(C(BPP)) = \sum_{i=1}^{n} w_i/c, \tag{8.13}$$

so a lower bound for BPP is

$$L_1 = \left\lceil \sum_{j=1}^{n} w_j / c \right\rceil . \tag{8.14}$$

Lower bound L_1 dominates the bound provided by the surrogate relaxation $S(BPP, \pi)$ given, for a positive vector (π_i) of multipliers, by

$$\text{minimize} \quad z = \sum_{i=1}^{n} y_i$$

$$\text{subject to} \quad \sum_{i=1}^{n} \pi_i \sum_{j=1}^{n} w_j x_{ij} \leq c \sum_{i=1}^{n} \pi_i y_i, \tag{8.15}$$

$$(8.3), (8.4), (8.5).$$

First note that we do not allow any multiplier, say $\pi_{\bar{\imath}}$, to take the value zero, since this would immediately produce a useless solution $x_{\bar{\imath}j} = 1$ for all $j \in N$. We then have the following

Theorem 8.1 *For any instance of BPP the optimal vector of multipliers for* $S(BPP, \pi)$ *is* $\pi_i = k$ *(k any positive constant) for all* $i \in N$.

Proof. Let $\bar{\imath} = \arg\min \{\pi_i : i \in N\}$, $\alpha = \pi_{\bar{\imath}}$, and suppose that (y_i^*) and (x_{ij}^*) define an optimal solution to $S(BPP, \pi)$. We can obtain a feasible solution of the same value by setting, for each $j \in N$, $x_{\bar{\imath}j}^* = 1$ and $x_{ij}^* = 0$ for $i \neq \bar{\imath}$. Hence $S(BPP, \pi)$ is equivalent to the problem

$$\text{minimize} \quad \sum_{i=1}^{n} y_i$$

$$\text{subject to} \quad \sum_{i=1}^{n} \pi_i y_i \geq \frac{\alpha}{c} \sum_{j=1}^{n} w_j,$$

$$y_i = 0 \text{ or } 1, \ i \in N,$$

i.e., to a special case of the 0-1 knapsack problem in minimization form, for which the optimal solution is trivially obtained by re-indexing the bins so that

$$\pi_1 \geq \pi_2 \geq \ldots \geq \pi_n \ (\equiv \alpha)$$

and setting $y_i = 1$ for $i \leq s = \min \{l \in N : \sum_{r=1}^{l} \pi_r \geq (\alpha/c) \sum_{j=1}^{n} w_j\}$, $y_i = 0$ for $i > s$. Hence the choice $\pi_i = \alpha$ (= k, any positive constant) for all $i \in N$ produces the maximum value of s, i.e. also of $z(S(BPP, \pi))$. □

Corollary 8.1 *When $\pi_i = k > 0$ for all $i \in N$, $z(S(BPP, \pi)) = z(C(BPP))$.*

Proof. With this choice of multipliers, $S(BPP, \pi)$ becomes

$$\text{minimize} \quad \sum_{i=1}^{n} y_i$$

$$\text{subject to} \quad \sum_{j=1}^{n} w_j \leq c \sum_{i=1}^{n} y_i,$$

$$y_i = 0 \text{ or } 1, \quad i \in N,$$

whose optimal solution value is $\sum_{j=1}^{n} w_j / c$. □

Lower bound L_1 also dominates the bound provided by the Lagrangian relaxation $L(BPP, \mu)$ defined, for a positive vector (μ_i) of multipliers, by

$$\text{minimize} \quad \sum_{i=1}^{n} y_i + \sum_{i=1}^{n} \mu_i \left(\sum_{j=1}^{n} w_j x_{ij} - c y_i \right) \tag{8.16}$$

$$\text{subject to (8.3), (8.4), (8.5).}$$

(Here again no multiplier of value zero can be accepted.)

Theorem 8.2 *For any instance of BPP the optimal choice of multipliers for $L(BPP, \mu)$ is $\mu_i = 1/c$ for all $i \in N$.*

Proof. We first prove that, given any vector (μ_i), we can obtain a better (higher) objective function value by setting, for all $i \in N$, $\mu_i = \mu_{\bar{\imath}}$, where $\bar{\imath} = \arg\min \{ \mu_i : i \in N \}$. In fact, by writing (8.16) as

$$\text{minimize} \quad \sum_{i=1}^{n} (1 - c\mu_i) y_i + \sum_{j=1}^{n} w_j \sum_{i=1}^{n} \mu_i x_{ij},$$

we see that the two terms can be optimized separately. The optimal (x_{ij}) values are clearly $x_{ij} = 0$ for $i \neq \bar{\imath}$ and $x_{\bar{\imath}j} = 1$, for all $j \in N$. It follows that, setting $\mu_i = \mu_{\bar{\imath}}$ for all $i \in N$, the first term is maximized, while the value of the second is unchanged.

Hence assume $\mu_i = k$ for all $i \in N$ (k any positive constant) and let us determine the optimal value for k. $L(BPP, \mu)$ becomes

$$\text{minimize} \quad \sum_{i=1}^{n} (1 - ck) y_i + k \sum_{j=1}^{n} w_j \tag{8.17}$$

$$\text{subject to} \quad y_i = 0 \text{ or } 1, \quad i \in N,$$

and its optimal solution is

(a) $y_i = 0$ for all $i \in N$, hence $z(L(BPP, \mu)) = k \sum_{j=1}^{n} w_j$, if $k \leq 1/c$;
(b) $y_i = 1$ for all $i \in N$, hence $z(L(BPP, \mu)) = n - k(cn - \sum_{j=1}^{n} w_j)$, if $k \geq 1/c$.

In both cases the highest value of the objective function $\sum_{j=1}^{n} w_j / c$ is provided by $k = 1/c$. □

Corollary 8.2 *When* $\mu_i = k = 1/c$ *for all* $i \in N$, $z(L(BPP, \mu)) = z(C(BPP))$.

Proof. Immediate from (8.17) and (8.13). □

A lower bound dominating L_1 can be obtained by dualizing in a Lagrangian fashion constraints (8.3). Given a vector (λ_j) of multipliers, the resulting relaxation, $L(BPP, \lambda)$, can be written as

$$\text{minimize} \quad \sum_{i=1}^{n} \left(y_i + \sum_{j=1}^{n} \lambda_j x_{ij} \right) - \sum_{j=1}^{n} \lambda_j \tag{8.18}$$

$$\text{subject to} \quad (8.2), (8.4), (8.5),$$

which immediately decomposes into n independent and identical problems (one for each bin). By observing that for any i, y_i will take the value 1 if and only if $x_{ij} = 1$ for at least one j, the optimal solution is obtained by defining

$$J^{<} = \{ j \in N : \lambda_j < 0 \}$$

and solving the 0-1 single knapsack problem

$$\text{maximize} \quad z(\lambda) = \sum_{j \in J^{<}} (-\lambda_j) q_j$$

$$\text{subject to} \quad \sum_{j \in J^{<}} w_j q_j \leq c,$$

$$q_j = 0 \text{ or } 1, \; j \in J^{<}.$$

If $z(\lambda) > 1$ then, for all $i \in N$, we have $y_i = 1$ and $x_{ij} = q_j$ (with $q_j = 0$ if $j \in N \backslash J^{<}$) for $j \in N$; otherwise we have $y_i = x_{ij} = 0$ for all $i, j \in N$. Hence

$$z(L(BPP, \lambda)) = \min\,(0, n(1 - z(\lambda))) - \sum_{j=1}^{n} \lambda_j .$$

It is now easy to see that, with the choice $\overline{\lambda}_j = -w_j/c$ for all $j \in N$, the resulting bound coincides with L_1. The objective function of the knapsack problem is in fact $(\sum_{j \in J^<} w_j q_j)/c$, with $J^< \equiv N$, so $z(\bar{\lambda}) \leq 1$ and $z(L(BPP, \overline{\lambda})) = \sum_{j=1}^{n} w_j/c = z(C(BPP))$.

Better multipliers can be obtained by using subgradient optimization techniques. Computational experiments, however, gave results worse than those obtained with the bounds described in the following sections.

8.3.2 A stronger lower bound

We first observe that the worst-case performance ratio of L_1 can easily be established as $r(L_1) = \frac{1}{2}$. Note, in fact, that in any optimal solution (x_{ij}) of value z, at most one bin (say the zth) can have $\sum_{j=1}^{n} w_j x_{zj} \leq c/2$ since, if two such bins existed, they could be replaced by a single bin. Hence $\sum_{j=1}^{n} w_j > \sum_{i=1}^{z-1} \sum_{j=1}^{n} w_j x_{ij} > (z-1)c/2$, from which $z \leq \lceil 2 \sum_{j=1}^{n} w_j/c \rceil$ and, from (8.14), $L_1/z \geq \frac{1}{2}$. To see that the ratio is tight, it is enough to consider the series of instances with $w_j = k+1$ for all $j \in N$ and $c = 2k$, for which $z = n$ and $L_1 = \lceil n(k+1)/2k \rceil$, so the ratio L_1/z can be arbitrarily close to $\frac{1}{2}$ for k sufficiently large.

Despite its simplicity, L_1 can be expected to have good average behaviour for problems where the weights are sufficiently small with respect to the capacity, since in such cases the evaluation is not greatly affected by the relaxation of the integrality constraints. For problems with larger weights, in which few items can be allocated, on average, to each bin, Martello and Toth (1990b) have proposed the following better bound.

Theorem 8.3 *Given any instance I of BPP, and any integer α, $0 \leq \alpha \leq c/2$, let*

$$J_1 = \{j \in N : w_j > c - \alpha\},$$

$$J_2 = \{j \in N : c - \alpha \geq w_j > c/2\},$$

$$J_3 = \{j \in N : c/2 \geq w_j \geq \alpha\};$$

then

$$L(\alpha) = |J_1| + |J_2| + \max\left(0, \left\lceil \frac{\sum_{j \in J_3} w_j - (|J_2|c - \sum_{j \in J_2} w_j)}{c} \right\rceil\right) \tag{8.19}$$

is a lower bound of $z(I)$.

Proof. Each item in $J_1 \cup J_2$ requires a separate bin, so $|J_1| + |J_2|$ bins are needed for them in any feasible solution. Let us relax the instance by replacing N with $(J_1 \cup J_2 \cup J_3)$. Because of the capacity constraint, no item in J_3 can be assigned to a bin containing an item of J_1. The total residual capacity of the $|J_2|$ bins needed for the items in J_2 is $\overline{c} = |J_2|c - \sum_{j \in J_2} w_j$. In the best case $\overline{c}$ will be completely filled by items in J_3, so the remaining total weight $\overline{w} = \sum_{j \in J_3} w_j - \overline{c}$, if any, will require $\lceil \overline{w}/c \rceil$ additional bins. □

Corollary 8.3 *Given any instance I of BPP,*

$$L_2 = \max \{L(\alpha) : 0 \leq \alpha \leq c/2, \ \alpha \ \textit{integer}\} \tag{8.20}$$

is a lower bound of $z(I)$.

Proof. Obvious. □

Lower bound L_2 dominates L_1. In fact, for any instance of BPP, using the value $\alpha = 0$, we have, from (8.19),

$$L(0) = 0 + |J_2| + \max\left(0, \left\lceil \frac{\sum_{j \in N} w_j - |J_2|c}{c} \right\rceil \right)$$

$$= |J_2| + \max(0, L_1 - |J_2|),$$

hence $L_2 \geq L(0) = \max(|J_2|, L_1)$.

Computing L_2 through (8.20) would require a pseudo-polynomial time. The same value, however, can be determined efficiently as follows.

Theorem 8.4 *Let V be the set of all the distinct values $w_j \leq c/2$. Then*

$$L_2 = \begin{cases} n & \textit{if } V = \emptyset; \\ \max\{L(\alpha) : \alpha \in V\} & \textit{otherwise}. \end{cases}$$

Proof. If $V = \emptyset$ the thesis is obvious from (8.19). Assuming $V \neq \emptyset$, we prove that, given $\alpha_1 < \alpha_2$, if α_1 and α_2 produce the same set J_3, then $L(\alpha_1) \leq L(\alpha_2)$. In fact: (a) the value $|J_1| + |J_2|$ is independent of α; (b) the value $(|J_2|c - \sum_{j \in J_2} w_j)$ produced by α_1 is no less than the corresponding value produced by α_2, since set J_2 produced by α_2 is a subset of set J_2 produced by α_1. Hence the thesis, since only distinct values $w_j \leq c/2$ produce, when used as α, different sets J_3, and each value w_j dominates the values $w_j - 1, \ldots, w_{j+1} + 1$ (by assuming that the weights satisfy (8.11)). □

Corollary 8.4 *If the items are sorted according to decreasing weights, L_2 can be computed in $O(n)$ time.*

Proof. Let

$$j^* = \min \{j \in N : w_j \le c/2\};$$

from Theorem 8.4, L_2 can be determined by computing $L(w_j)$ for $j = j^*, j^* + 1, \dots, n$, by considering only distinct w_j values. The computation of $L(w_{j^*})$ clearly requires $O(n)$ time. Since $|J_1| + |J_2|$ is a constant, the computation of each new $L(w_j)$ simply requires to update $|J_2|$, $\sum_{j \in J_3} w_j$ and $\sum_{j \in J_2} w_j$. Hence all the updatings can be computed in $O(n)$ time since they correspond to a constant time for each $j = j^* + 1, \dots, n$. □

The average efficiency of the above computation can be improved as follows. At any iteration, let L_2^* be the largest $L(w_j)$ value computed so far. If $|J_1| + |J_2| + \lceil (\sum_{j=j^*}^{n} w_j - (|J_2|c - \sum_{j \in J_2} w_j))/c \rceil \le L_2^*$, then (see point (b) in the proof of Theorem 8.4) no further iteration could produce a better bound, so $L_2 = L_2^*$.

Example 8.1

Consider the instance of BPP defined by

$$n = 9,$$

$$(w_j) = (70, 60, 50, 33, 33, 33, 11, 7, 3),$$

$$c = 100.$$

An optimal solution requires 4 bins for item sets $\{1, 7, 8, 9\}$, $\{2, 4\}$, $\{3, 5\}$ and $\{6\}$, respectively.

From (8.14),

$$L_1 = \lceil 300/100 \rceil = 3.$$

In order to determine L_2 we compute, using (8.19) and Corollary 8.4,

$$L(50) = 2 + 0 + \max(0, \lceil (50 - 0)/100 \rceil) = 3;$$

$$L(33) = 1 + 1 + \max(0, \lceil (149 - 40)/100 \rceil) = 4;$$

since at this point we have $1 + 1 + \lceil (170 - 40)/100 \rceil = 4$, the computation can be terminated with $L_2 = 4$. □

The following procedure efficiently computes L_2. It is assumed that, on input, the items are sorted according to (8.11) and $w_n \le c/2$. (If $w_n > c/2$ then, trivially, $L_2 = n = z$.) Figure 8.1 illustrates the meaning of the main variables of the procedure.

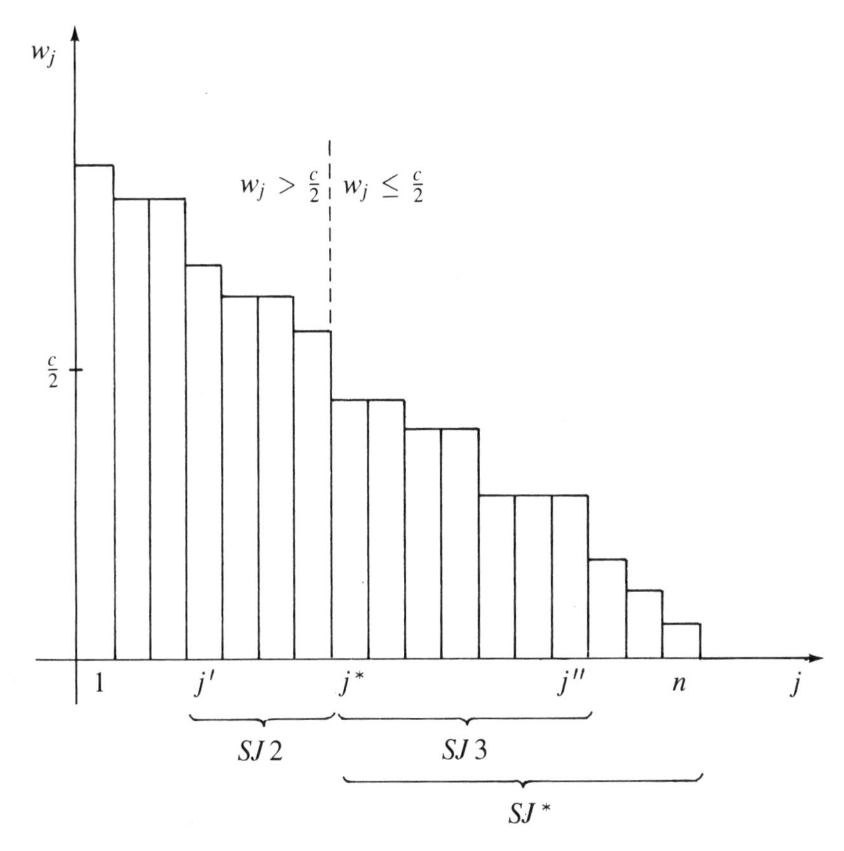

Figure 8.1 Main variables in procedure L2

procedure L2:
input: $n, (w_j), c$;
output: L_2;
begin
$\quad N := \{1, \dots, n\}$;
$\quad j^* :=$ min$\{j \in N : w_j \leq c/2\}$;
$\quad$ **if** $j^* = 1$ **then** $L_2 := \lceil \sum_{j=1}^{n} w_j / c \rceil$
$\quad$ **else**
$\quad\quad$ **begin**
$\quad\quad\quad CJ12 := j^* - 1$ (**comment** : $CJ12 = |J_1| + |J_2|$);
$\quad\quad\quad SJ^* := \sum_{j=j^*}^{n} w_j$;
$\quad\quad\quad j' :=$ min$\{j \in N : j < j^*$ and $w_j \leq c - w_{j^*}\}$($j' := j^*$ if no such w_j);
$\quad\quad\quad CJ2 := j^* - j'$ (**comment** : $CJ2 = |J_2|$) ;
$\quad\quad\quad SJ2 := \sum_{j=j'}^{j^*-1} w_j$ (**comment** : $SJ2 = \sum_{j \in J_2} w_j$);
$\quad\quad\quad j'' := j^*$;
$\quad\quad\quad SJ3 := w_{j''}$;
$\quad\quad\quad w_{n+1} := 0$;
$\quad\quad\quad$ **while** $w_{j''+1} = w_{j''}$ **do**

```
                begin
                      j'' := j'' + 1;
                      SJ3 := SJ3 + w_{j''}
                end (comment : SJ3 = Σ_{j∈J_3} w_j);
          L_2 := CJ12;
          repeat
                L_2 := max(L_2, CJ12 + ⌈(SJ3 + SJ2)/c − CJ2⌉);
                j'' := j'' + 1;
                if j'' ≤ n then
                      begin
                            SJ3 := SJ3 + w_{j''};
                            while w_{j''+1} = w_{j''} do
                                  begin
                                        j'' := j'' + 1;
                                        SJ3 := SJ3 + w_{j''}
                                  end;
                            while j' > 1 and w_{j'−1} ≤ c − w_{j''} do
                                  begin
                                        j' := j' − 1;
                                        CJ2 := CJ2 + 1;
                                        SJ2 := SJ2 + w_{j'}
                                  end
                      end
          until j'' > n or CJ12 + ⌈(SJ* + SJ2)/c − CJ2⌉ ≤ L_2
    end
end.
```

The worst-case performance ratio of L_2 is established by the following

Theorem 8.5 $r(L_2) = \frac{2}{3}$.

Proof. Let I be any instance of BPP and z its optimal solution value. We prove that $L_2 \geq L(0) \geq \frac{2}{3}z$. Hence, let $\alpha = 0$, i.e. $J_1 = \emptyset$, $J_2 = \{j \in N : w_j > c/2\}$, $J_3 = N \backslash J_2$. If $J_3 = \emptyset$, then, from (8.19), $L(0) = |J_2| = n = z$. Hence assume $J_3 \neq \emptyset$. Let $\overline{I}$ denote the instance we obtain by relaxing the integrality constraints on x_{ij} for all $j \in J_3$ and $i \in N$. It is clear that $L(0)$ is the value of the optimal solution to $\overline{I}$, which can be obtained as follows. $|J_2|$ bins are first initialized for the items in J_2. Then, for each item $j \in J_3$, let i^* denote the lowest indexed bin not completely filled (if no such bin, initialize a new one) and $c(i^*) \leq c$ its residual capacity. If $w_j \leq c(i^*)$ then item j is assigned to bin i^*; otherwise item j is replaced by two items j_1, j_2 with $w_{j_1} = c(i^*)$ and $w_{j_2} = w_j - w_{j_1}$, item j_1 is assigned to bin i^* and the process is continued with item j_2. In this solution $L(0) - 1$ items at most are split (no splitting can occur in the $L(0)$th bin). We can now obtain a feasible solution of value $\tilde{z} \geq z$ to I by removing the split items from the previous solution and assigning them to new bins. By the definition of J_3, at most $\lceil (L(0) - 1)/2 \rceil$ new bins are needed, so $\tilde{z} \leq L(0) + \lfloor L(0)/2 \rfloor$, hence $\frac{3}{2}$ $L(0) \geq z$.

To prove that the ratio is tight, consider the series of instances with n even,

$w_j = k+1$ ($k \geq 2$) for $j = 1, \dots, n$ and $c = 3k$. We have $z = n/2$ and $L_2 = L(k+1) = \lceil n(k+1)/(3k) \rceil$, so ratio L_2/z can be arbitrarily close to $\frac{2}{3}$ for k sufficiently large. □

It is worthy of note that lower bounds with better worst-case performance can easily be obtained from approximate algorithms. We can use, for example, algorithm BFD of Section 8.2 to produce, for any instance I, a solution of value $BFD(I)$. This solution (see Johnson, Demers, Ullman, Garey and Graham (1974)) satisfies the same worst-case bound as $FFD(I)$, so we trivially obtain a lower bound (see (8.12))

$$LBFD(I) = \frac{9}{11}(BFD(I) - 4), \tag{8.21}$$

whose worst-case performance is smaller than that of L_2 for $z(I)$ sufficiently large, and asymptotically tends to $\frac{9}{11}$. Since however $BFD(I)$ is known to be, in general, close to $z(I)$, the average performance of $LBFD$ is quite poor (as will be seen in Section 8.6).

8.4 REDUCTION ALGORITHMS

The reduction techniques described in the present section are based on the following dominance criterion (Martello and Toth, 1990b).

We define a *feasible set* as a subset $F \subseteq N$ such that $\sum_{j \in F} w_j \leq c$. Given two feasible sets F_1 and F_2, we say that F_1 *dominates* F_2 if the value of the optimal solution which can be obtained by imposing for a bin, say i^*, the values $x_{i^*j} = 1$ if $j \in F_1$ and $x_{i^*j} = 0$ if $j \notin F_1$, is no greater than the value that can be obtained by forcing the values $x_{i^*j} = 1$ if $j \in F_2$ and $x_{i^*j} = 0$ if $j \notin F_2$. A possible way to check such situations is the following

Dominance Criterion *Given two distinct feasible sets F_1 and F_2, if a partition of F_2 into subsets $P_1, \dots, P_l$ and a subset $\{j_1, \dots, j_l\}$ of F_1 exist such that $w_{j_h} \geq \sum_{k \in P_h} w_k$ for $h = 1, \dots, l$, then F_1 dominates F_2.*

Proof. Completing the solution through assignment of the items in $N \backslash F_1$ is easier than through assignment of the items in $N \backslash F_2$. In fact: (a) $\sum_{j \in N \backslash F_1} w_j \leq \sum_{j \in N \backslash F_2} w_j$; (b) for any feasible assignment of an item $j_h \in \{j_1, \dots, j_l\} \subseteq F_1$ there exists a feasible assignment of the items in $P_h \subseteq F_2$ (while the opposite does not hold). □

If a feasible set F dominates all the others, then the items of F can be assigned to a bin and removed from N. Checking all such situations, however, is clearly impractical. The following algorithm limits the search to sets of cardinality not greater than 3 and avoids the enumeration of useless sets. It considers the items according to decreasing weights and, for each item j, it checks for the existence of a feasible set F such that $j \in F$, with $|F| \leq 3$, dominating all feasible sets containing item j. Whenever such a set is found, the corresponding items are assigned to a new bin and the search continues with the remaining items. It is assumed that, on input, the items are sorted according to (8.11). On output, z^r gives the number of optimally filled bins, and, for each $j \in N$,

$$b_j = \begin{cases} 0 & \text{if item } j \text{ has not been assigned;} \\ \text{bin to which it has been assigned, otherwise.} \end{cases}$$

```
procedure MTRP:
input: n, (w_j), c;
output: z^r, (b_j);
begin
    N := {1,...,n};
    N̄ := Ø;
    z^r := 0;
    for j := 1 to n do b_j := 0;
    repeat
        find j = min{h : h ∈ N\N̄};
        let N' = N\{j} = {j_1,...,j_l} with w_{j_1} ≥ ... ≥ w_{j_l};
        F := Ø;
        find the largest k such that w_j + Σ_{q=l-k+1}^{l} w_{j_q} ≤ c;
        if k = 0 then F := {j}
        else
            begin
                j* := min {h ∈ N' : w_j + w_h ≤ c};
                if k = 1 or w_j + w_{j*} = c then F := {j, j*}
                else if k = 2 then
                    begin
                        find j_a, j_b ∈ N', with a < b, such that
                            w_{j_a} + w_{j_b} = max {w_{j_r} + w_{j_s} :
                            j_r, j_s ∈ N', w_j + w_{j_r} + w_{j_s} ≤ c};
                        if w_{j*} ≥ w_{j_a} + w_{j_b} then F := {j, j*}
                        else if w_{j*} = w_{j_a} and (b − a ≤ 2
                                or w_j + w_{j_b−1} + w_{j_b−2} > c)
                            then F := {j, j_a, j_b}
                    end
            end;
        if F = Ø then N̄ := N̄ ∪ {j}
        else
            begin
                z^r := z^r + 1;
                for each h ∈ F do b_h = z^r;
                N := N\F
            end
    until N\N̄ = Ø
end.
```

At each iteration, $k+1$ gives the maximum cardinality of a feasible set containing item j. Hence it immediately follows from the dominance criterion that $F = \{j\}$ when $k = 0$, and $F = \{j, j^*\}$ when $k = 1$ or $w_j + w_{j^*} = c$. When $k = 2$, (a) if $w_{j^*} \geq w_{j_a} + w_{j_b}$ then set $\{j^*\}$ dominates all pairs of items (and, by definition of j^*, all singletons) which can be packed together with j, so $\{j, j^*\}$ dominates all feasible sets containing j; (b) if $w_{j^*} = w_{j_a}$ and either $b - a \leq 2$ or $w_j + w_{j_b - 1} + w_{j_b - 2} > c$ then set $\{j_a, j_b\}$ dominates all pairs and all singletons which can be packed together with j.

The time complexity of MTRP is $O(n^2)$. In fact, the repeat-until loop is executed $O(n)$ times. At each iteration, the heaviest step is the determination of j_a and j_b, which can easily be implemented so as to require $O(n)$ time, since the pointers r and s (assuming $r < s$) must be moved only from left to right and from right to left, respectively.

The reduction procedure above can also be used to determine a new lower bound L_3. After execution of procedure MTRP for an instance I of BPP, let z_1^r denote the output value of z^r, and $I(z_1^r)$ the corresponding residual instance, defined by item set $\{j \in N : b_j = 0\}$. It is obvious that a lower bound for I is given by $z_1^r + L(I(z_1^r))$, where $L(I(z_1^r))$ denotes the value of any lower bound for $I(z_1^r)$. (Note that $z_1^r + L(I(z_1^r)) \geq L(I)$.) Suppose now that $I(z_1^r)$ is relaxed in some way (see below) and MTRP is applied to the relaxed instance, producing the output value z_2^r and a residual relaxed instance $I(z_1^r, z_2^r)$. A lower bound for I is then $z_1^r + z_2^r + L(I(z_1^r, z_2^r))$. Iterating the process we obtain a series of lower bounds of the form

$$L_3 = z_1^r + z_2^r + \ldots + L(I(z_1^r, z_2^r, \ldots)).$$

The following procedure computes the maximum of the above bounds, using L_2 for L. At each iteration, the current residual instance is relaxed through removal of the smallest item. It is assumed that on input the items are sorted according to (8.11).

```
procedure L3:
input: n, (w_j), c;
output: L3;
begin
    L3 := 0;
    z := 0;
    n̄ := n;
    for j := 1 to n do w̄_j := w_j;
    while n̄ ≥ 1 do
        begin
            call MTRP giving n̄, (w̄_j) and c, yielding z^r and (b̄_j);
            z := z + z^r;
            k := 0;
            for j := 1 to n̄ do
                if b̄_j = 0 then
```

```
            begin
                k := k + 1;
                w̄_k := w̄_j
            end;
        n̄ := k;
        if n̄ = 0 then L2 := 0
        else call L2 giving n̄, (w̄_j) and c, yielding L2;
        L3 := max(L3, z + L2);
        n̄ := n̄ − 1 (comment: removal of the smallest item)
    end
end.
```

Since MTRP runs in $O(n^2)$ time, the overall time complexity of L3 is $O(n^3)$. It is clear that $L_3 \geq L_2$.

Note that only the reduction determined in the first iteration of MTRP is valid for the original instance, since the other reductions are obtained after one or more relaxations. If however, after the execution of L3, *all* the removed items can be assigned to the bins filled up by the executions of MTRP, then we obtain a feasible solution of value L_3, i.e. optimal.

Example 8.2

Consider the instance of BPP defined by

$n = 14,$

$(w_j) = (99, 94, 79, 64, 50, 46, 43, 37, 32, 19, 18, 7, 6, 3),$

$c = 100.$

The first execution of MTRP gives

$j = 1 : k = 0, F = \{1\};$

$j = 2 : k = 1, j^* = 13, F = \{2, 13\},$

and $F = \emptyset$ for $j \geq 3$. Hence

$z = 2; (\overline{b}_j) = (1, 2, 0, 0, 0, 0, 0, 0, 0, 0, 0, 0, 2, 0);$

executing L2 for the residual instance we get $L_2 = 4$, so

$L_3 = 6.$

Item 14 is now removed and MTRP is applied to item set $\{3, 4, \ldots, 12\}$, producing (indices refer to the original instance)

$j = 3 : k = 1, j^* = 10, F = \{3, 10\};$

$j = 4 : k = 2, j^* = 9, j_a = 11, j_b = 12, F = \{4, 9\};$

$j = 5 : k = 2, j^* = 6, j_a = 7, j_b = 12, F = \emptyset;$

$j = 6 : k = 2, j^* = 5, j_a = 7, j_b = 12, F = \{6, 5\};$

$j = 7 : k = 2, j^* = 8, j_a = 8, j_b = 11, F = \{7, 8, 11\};$

$j = 12 : k = 0, F = \{12\};$

numbering the new bins with 3, 4, ..., 7 we thus obtain

$$z = 7; \ (\overline{b}_j) = (1, 2, 3, 4, 5, 5, 6, 6, 4, 3, 6, 7, 2, -\ );$$

hence $L_2 = 0$ (since $\overline{n} = 0$) and the execution terminates with $L_3 = 7$.

Noting now that the eliminated item 14 can be assigned, for example to bin 4, we conclude that all reductions are valid for the original instance. The solution obtained (with $\overline{b}_{14} = 4$) is also optimal, since all items are assigned. □

8.5 EXACT ALGORITHMS

As already mentioned, very little can be found in the literature on the exact solution of BPP.

Eilon and Christofides (1971) have presented a simple depth-first enumerative algorithm based on the following "best-fit decreasing" branching strategy. At any decision node, assuming that b bins have been initialized, let $(\overline{c}_{i_1}, \ldots, \overline{c}_{i_b})$ denote their current residual capacities sorted by increasing value, and $\overline{c}_{i_{b+1}} \equiv c_{b+1} = c$ the capacity of the next (not yet initialized) bin: the branching phase assigns the free item j^* of largest weight, in turn, to bins $i_s, \ldots, i_b, i_{b+1}$, where $s = \min \{h : 1 \leq h \leq b+1, \ \ \overline{c}_{i_h} + w_{j^*} \leq c\}$. Lower bound L_1 (see Section 8.3.1) is used to fathom decision nodes.

Hung and Brown (1978) have presented a branch-and-bound algorithm for a generalization of BPP to the case in which the bins are allowed to have different capacities. Their branching strategy is based on a characterization of equivalent assignments, which reduces the number of explored decision nodes. The lower bound employed is again L_1.

We do not give further details on these algorithms, since the computational results reported in Eilon and Christofides (1971) and Hung and Brown (1978) indicate that they can solve only small-size instances.

Martello and Toth (1989) have proposed an algorithm, MTP, based on a "first-fit decreasing" branching strategy. The items are initially sorted according to decreasing weights. The algorithm indexes the bins according to the order in which they are initialized. At each decision node, the first (i.e. largest) free item is assigned, in turn, to the feasible initialized bins (by increasing index) and to a new bin. At any *forward step*, (a) procedures L_2 and then L_3 are called to attempt to fathom the node and reduce the current problem; (b) when no fathoming occurs, approximate algorithms FFD, BFD (see Section 8.2) and WFD are applied to the current problem, to try and improve the best solution so far. (A *Worst-Fit Decreasing* (WFD) approximate algorithm for BPP sorts the items by decreasing weights and assigns each item to the feasible initialized bin (if any) of largest residual capacity.) A *backtracking step* implies the removal of the current item j^* from its current bin i^*, and its assignment to the next feasible bin (but backtracking occurs if i^* had been initialized by j^*, since initializing i^*+1 with j^* would produce an identical situation). If z is the value of the current optimal solution, whenever backtracking must occur, it is performed on the last item assigned to a bin of index not greater than $z - 2$ (since backtracking on any item assigned to bin z or $z - 1$ would produce solutions requiring at least z bins).

In addition, the following *dominance criterion* between decision nodes is used. When the current item j^* is assigned to a bin i^* whose residual capacity $\overline{c}_{i^*}$ is less than $w_{j^*} + w_n$, this assignment dominates all the assignments to i^* of items $j > j^*$ which do not allow the insertion of at least one further item. Hence such assignment "closes" bin i^*, in the sense that, after backtracking on j^*, no item $j \in \{k > j^* : w_k + w_n > \overline{c}_{i^*}\}$ is assigned to i^*; the bin is "re-opened" when the first item $j > j^*$ for which $w_j + w_n \leq \overline{c}_{i^*}$ is considered or, if no such item exists, when the first backtracking on an item $l < j^*$ is performed.

Since at any decision node the current residual capacities $\overline{c}_i$ of the bins are different, the computation of lower bounds L_2 and L_3 must take into account this situation. An easy way is to relax the current instance by adding one extra item of weight $c - \overline{c}_i$ to the free items for each initialized bin i, and by supposing that all the bins have capacity c.

Example 8.3

Consider the instance of MTP defined by

$$n = 10;$$

$$(w_j) = (49, 41, 34, 33, 29, 26, 26, 22, 20, 19);$$

$$c = 100.$$

We define a feasible solution through vector (b_j), with

$$b_j = \text{bin to which item } j \text{ is assigned } (j = 1, \ldots, n);$$

Figure 8.2 gives the decision-tree produced by algorithm MTP. Initially, all lower bound computations give the value 3, while approximate algorithm FFD gives the first feasible solution

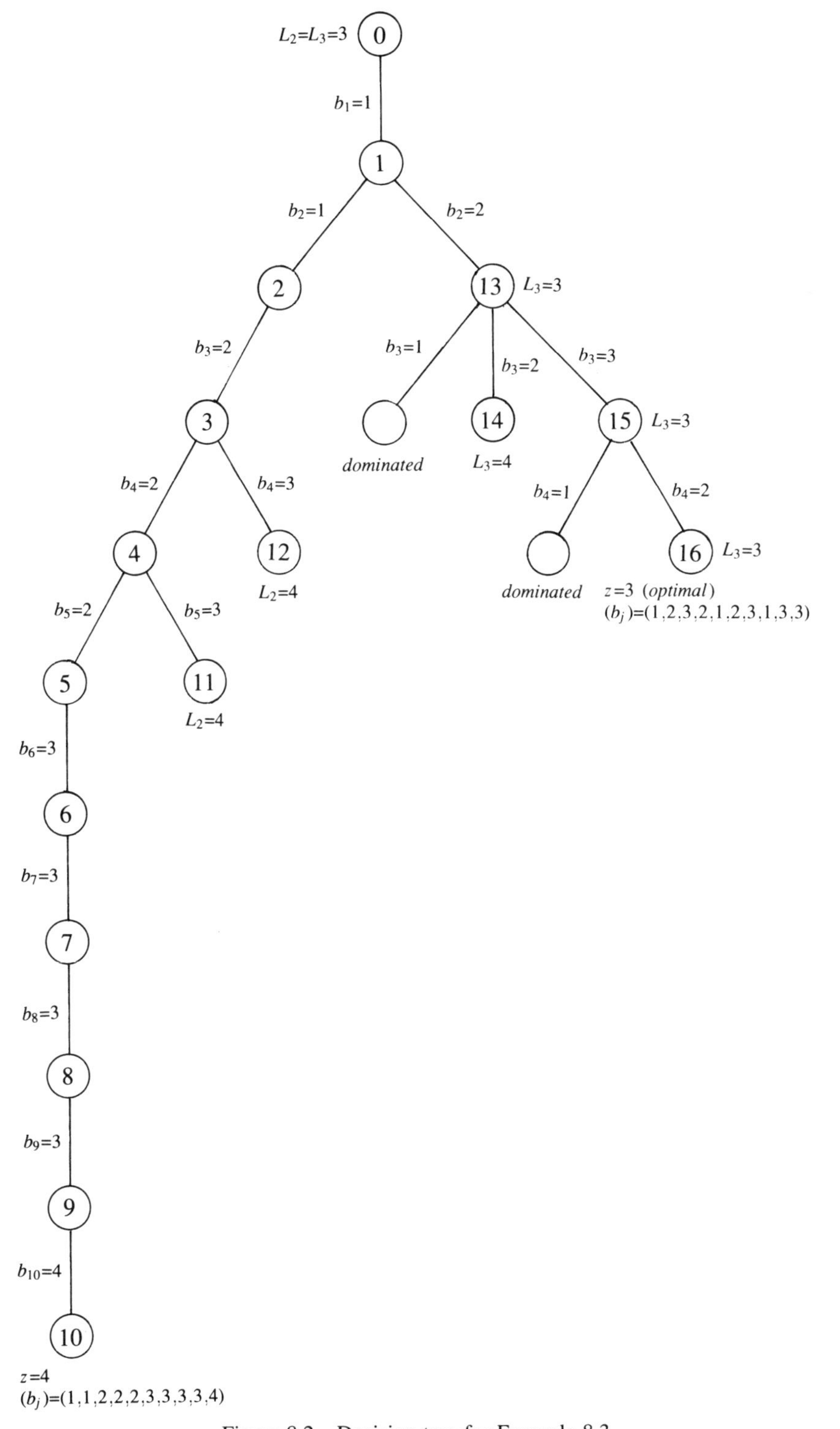

Figure 8.2 Decision-tree for Example 8.3

$$z = 4,$$

$$(b_j) = (1, 1, 2, 2, 2, 3, 3, 3, 3, 4),$$

corresponding to decision-nodes 1–10. No second son is generated by nodes 5-9, since this would produce a solution of value 4 or more. Nodes 11 and 12 are fathomed by lower bound L_2. The first son of node 2 initializes bin 2, so no further son is generated. The first son of node 13 is dominated by node 2, since in both situations no further item can be assigned to bin 1; for the same reason node 2 dominates the first son of node 15. Node 14 is fathomed by lower bound L_3. At node 16, procedure MTRP (called by L3) is applied to problem

$$\overline{n} = 9,$$

$$(\overline{w}_j) = (74, 49, 34, 29, 26, 26, 22, 20, 19),$$

$$\overline{c} = 100,$$

and optimally assigns to bin 2 the first and fifth of these items (corresponding to items 2, 4 and 6 of the original instance). Then, by executing the approximate algorithm FFD for the reduced instance

$$(w_j) = (-, -, -, -, 29, -, 26, 22, 20, 19),$$

$$(\overline{c}_i) = (51, 0, 66, 100, 100, \ldots),$$

where $\overline{c}_i$ denotes the residual capacity of bin i, we obtain

$$(\overline{b}_j) = (-, -, -, -, 1, -, 3, 1, 3, 3),$$

hence an overall solution of value 3, i.e. optimal. □

The Fortran implementation of algorithm MTP is included in the present volume.

8.6 COMPUTATIONAL EXPERIMENTS

In this section we examine the average computational performance of the lower bounds (Sections 8.3–8.4) and of the exact algorithm MTP (Section 8.5). The procedures have been coded in Fortran IV and run on an HP 9000/840 (using option "-o" for the compiler) on three classes of randomly generated item sizes:

Class 1: w_j uniformly random in [1, 100];

Class 2: w_j uniformly random in [20, 100];

Class 3: w_j uniformly random in [50, 100].

Table 8.2 $C = 100$. HP 9000/840 in seconds. Average times / Average percentage errors (exact solution values found) over 20 problems

		LBFD		L_1		L_2		L_3	
Class	n	Time	% err(opt)	Time	% err(opt)	Time	% err(opt)	Time	% err(opt)
1	50	0.003	28.075(0)	0.001	6.413(2)	0.001	0.519(17)	0.001	0.000(20)
	100	0.009	24.530(0)	0.001	3.549(0)	0.001	0.877(11)	0.006	0.196(18)
	200	0.020	20.934(0)	0.002	2.539(1)	0.001	0.291(14)	0.013	0.149(17)
	500	0.061	19.175(0)	0.002	2.487(0)	0.003	0.194(12)	0.033	0.078(16)
	1 000	0.136	18.452(0)	0.003	1.127(0)	0.007	0.159(7)	0.091	0.060(15)
2	50	0.004	27.814(0)	0.001	7.938(0)	0.001	0.308(18)	0.002	0.000(20)
	100	0.011	23.641(0)	0.001	6.846(0)	0.002	0.306(16)	0.003	0.000(20)
	200	0.022	20.192(0)	0.001	6.084(0)	0.001	0.352(13)	0.009	0.039(19)
	500	0.059	19.506(0)	0.001	5.558(0)	0.003	0.189(10)	0.019	0.032(18)
	1 000	0.139	18.441(0)	0.002	4.805(0)	0.005	0.168(5)	0.058	0.032(16)
3	50	0.004	24.281(0)	0.001	23.300(0)	0.001	0.000(20)	0.001	0.000(20)
	100	0.010	21.612(0)	0.001	24.012(0)	0.001	0.000(20)	0.003	0.000(20)
	200	0.023	19.483(0)	0.001	24.003(0)	0.001	0.000(20)	0.005	0.000(20)
	500	0.062	18.841(0)	0.001	24.044(0)	0.001	0.000(20)	0.016	0.000(20)
	1 000	0.146	18.153(0)	0.002	24.285(0)	0.001	0.000(20)	0.038	0.000(20)

Table 8.3 $C = 120$. HP 9000/840 in seconds. Average times / Average percentage errors (exact solution values found) over 20 problems

		LBFD		L_1		L_2		L_3	
Class	n	Time	% err(opt)	Time	% err(opt)	Time	% err(opt)	Time	% err(opt)
	50	0.004	29.832(0)	0.001	3.788(10)	0.001	0.000(20)	0.005	0.000(20)
	100	0.008	24.562(0)	0.001	1.990(10)	0.002	0.474(16)	0.014	0.357(17)
1	200	0.020	21.180(0)	0.001	0.732(13)	0.003	0.165(17)	0.042	0.000(20)
	500	0.055	19.731(0)	0.001	0.538(13)	0.006	0.116(15)	0.232	0.047(18)
	1 000	0.126	18.573(0)	0.002	0.105(17)	0.008	0.035(17)	0.983	0.024(18)
	50	0.003	28.514(0)	0.001	7.924(0)	0.001	0.976(15)	0.003	0.408(18)
	100	0.009	23.881(0)	0.001	3.714(2)	0.001	0.478(15)	0.005	0.091(19)
2	200	0.019	20.212(0)	0.001	3.326(0)	0.001	0.580(10)	0.009	0.050(19)
	500	0.060	19.304(0)	0.001	3.106(0)	0.004	0.216(11)	0.023	0.059(17)
	1 000	0.126	18.607(0)	0.003	1.386(0)	0.007	0.239(8)	0.062	0.080(13)
	50	0.004	25.208(0)	0.001	24.145(0)	0.001	0.000(20)	0.002	0.000(20)
	100	0.010	21.503(0)	0.001	22.344(0)	0.001	0.000(20)	0.004	0.000(20)
3	200	0.023	19.752(0)	0.001	22.155(0)	0.001	0.063(18)	0.007	0.000(20)
	500	0.064	18.614(0)	0.002	22.035(0)	0.002	0.012(19)	0.017	0.000(20)
	1 000	0.147	18.323(0)	0.003	21.775(0)	0.003	0.056(12)	0.036	0.000(20)

Table 8.4 $C = 150$. HP 9000/840 in seconds. Average times / Average percentage errors (exact solution values found) over 20 problems

		LBFD		L_1		L_2		L_3	
Class	n	Time	% err(opt)	Time	% err(opt)	Time	% err(opt)	Time	% err(opt)
1	50	0.003	31.832(0)	0.001	0.488(18)	0.001	0.250(19)	0.011	0.250(19)
	100	0.009	25.634(0)	0.001	0.000(20)	0.003	0.000(20)	0.039	0.000(20)
	200	0.016	21.984(0)	0.001	0.000(20)	0.002	0.000(20)	0.137	0.000(20)
	500	0.056	20.007(0)	0.002	0.000(20)	0.004	0.000(20)	0.857	0.000(20)
	1 000	0.119	19.189(0)	0.003	0.000(20)	0.009	0.000(20)	4.011	0.000(20)
2	50	0.004	31.408(0)	0.001	0.617(18)	0.001	0.217(19)	0.011	0.217(19)
	100	0.009	23.652(0)	0.001	0.727(14)	0.001	0.727(14)	0.043	0.727(14)
	200	0.020	20.984(0)	0.001	0.978(0)	0.004	0.978(5)	0.169	0.978(5)
	500	0.059	19.837(0)	0.002	1.153(0)	0.004	1.153(0)	1.139	1.153(0)
	1 000	0.128	18.416(0)	0.003	1.224(0)	0.005	1.224(0)	9.806	1.224(0)
3	50	0.003	27.941(0)	0.001	8.851(1)	0.001	0.579(17)	0.002	0.000(20)
	100	0.009	23.247(0)	0.001	4.302(4)	0.002	0.665(13)	0.003	0.000(20)
	200	0.021	20.671(0)	0.001	3.856(0)	0.002	0.625(9)	0.007	0.000(20)
	500	0.061	19.459(0)	0.001	3.480(0)	0.003	0.234(11)	0.019	0.000(20)
	1 000	0.132	18.527(0)	0.003	1.687(0)	0.005	0.328(8)	0.036	0.000(20)

For each class, three values of c have been considered: $c = 100$, $c = 120$, $c = 150$. For each pair (class, value of c) and for different values of n ($n =$ 50, 100, 200, 500, 1000), 20 instances have been generated.

In Tables 8.2–8.4 we examine the behaviour of lower bounds LBFD, L1, L2 and L3. The entries give, for each bound, the average computing time (expressed in seconds and not comprehensive of the sorting time), the average percentage error and, in brackets, the number of times the value of the lower bound coincided with that of the optimal solution. LBFD requires times almost independent of the data generation and, because of the good approximation produced by the best-fit decreasing algorithm, gives high errors, tending to $\frac{2}{11}$ when n grows. L_1 obviously requires very small times, practically independent of the data generation; the tightness improves when the ratio $c/\min_j\{w_j\}$ grows, since the computation is based on continuous relaxation of the problem. L_2 requires slightly higher times, but produces tighter values; for class 1 it improves when c grows, for classes 2 and 3 it get worse when c grows. The times required by L_3 are in general comparatively very high (because of the iterated execution of reduction procedure MTRP), and clearly grow both with n and c; the approximation produced is generally very good, with few exceptions.

Note that the problems generated can be considered "hard", since few items are packed in each bin. Using the value $c = 1000$, L_1 requires the same times and almost always produces the optimal solution value.

Table 8.5 gives the results obtained by the exact algorithm MTP for the instances used for the previous tables. The entries give average running time (expressed in seconds and comprehensive of the sorting time) and average number of nodes

Table 8.5 Algorithm MTP. HP 9000/840 in seconds. Average times/Average numbers of nodes over 20 problems

		$c = 100$		$c = 120$		$c = 150$	
Class	n	Time	Nodes	Time	Nodes	Time	Nodes
1	50	0.006	0	0.005	0	0.096	11
	100	0.012	1	15.022(17)	3561	0.156	29
	200	5.391	1114	0.062	6	0.140	10
	500	10.236	2805	10.340	887	2.124	28
	1000	20.206(16)	2686	6.596	244	8.958	44
2	50	0.005	0	0.008	1	0.183	61
	100	0.012	1	0.030	9	26.599(15)	4275
	200	0.047	11	0.073	18	69.438(7)	8685
	500	0.127	28	10.062	1663	—	—
	1000	15.524(17)	3896	30.148(14)	4774	—	—
3	50	0.005	0	0.005	0	0.005	0
	100	0.010	0	0.010	0	0.010	0
	200	0.019	0	0.020	0	0.018	0
	500	0.049	0	0.050	0	0.051	0
	1000	0.102	0	0.104	0	0.105	0

explored in the branch-decision tree. A time limit of 100 seconds was assigned to the algorithm for each problem instance. When the time limit occurred, the corresponding entry gives, in brackets, the number of instances solved to optimality (the average values are computed by also considering the interrupted instances). When less than half of the 20 instances generated for an entry was completed, larger values of n were not considered.

All the instances of Class 3 were solved very quickly, since procedure L3 always produced the optimal solution. For Class 1 the results are very satisfactory, with few exceptions. On Class 2, the behaviour of the algorithm was better than on Class 1 for $c = 100$, about the same for $c = 120$, and clearly worse for $c = 150$. Worth noting is that in only a few cases the optimal solution was found by the approximate algorithms used.

Appendix: Computer codes

A.1 INTRODUCTION

The diskette included in the volume contains the Fortran implementations of the most effective algorithms described in the various chapters. Table A.1 gives, for each code, the problem solved, the approximate number of lines (including comments), the section where the corresponding procedure (which has the same name as the code) is described, and the type of algorithm implemented. Most of the implementations are exact branch-and-bound algorithms which can also be used to provide approximate solutions by limiting the number of backtrackings through an input parameter (notation Exact/Approximate in the table).

Table A.1 Fortran codes included in the volume

Code	Problem	Lines	Section	Type of algorithm
MT1	0-1 Knapsack	280	2.5.2	Exact
MT1R	0-1 Knapsack	300	2.5.2	Exact (real data)
MT2	0-1 Knapsack	1400	2.9.3	Exact/Approximate
MTB2	Bounded Knapsack	190 (+1400)*	3.4.2	Exact/Approximate
MTU2	Unbounded Knapsack	1100	3.6.3	Exact/Approximate
MTSL	Subset-Sum	780	4.2.3	Exact/Approximate
MTC2	Change-Making	450	5.6	Exact/Approximate
MTCB	Bounded Change-Making	380	5.8	Exact/Approximate
MTM	0-1 Multiple Knapsack	670	6.4.3	Exact/Approximate
MTHM	0-1 Multiple Knapsack	590	6.6.2	Approximate
MTG	Generalized Assignment	2300	7.3	Exact/Approximate
MTHG	Generalized Assignment	500	7.4	Approximate
MTP	Bin Packing	1330	8.5	Exact/Approximate

* MTB2 must be linked with MT2.

All programs solve problems defined by integer parameters, except MT1R which solves the 0-1 single knapsack problem with real parameters.

All codes are written according to PFORT, a portable subset of 1966 ANSI Fortran, and are accepted by the PFORT verifier developed by Ryder and Hall (1981) at Bell Laboratories. The codes have been tested on a Digital VAX 11/780 and a Hewlett-Packard 9000/840.

With the only exception of MTB2 (which must be linked with MT2), the codes are completely self-contained. Communication to the codes is achieved solely through the parameter list of a "main" subroutine whose name is that of the code.

The following sections give, for each problem and for each code, the corresponding comment and specification statements.

A.2 0-1 KNAPSACK PROBLEM

A.2.1 Code MT1

SUBROUTINE MT1 (N, P, W, C, Z, X, JDIM, JCK,
XX, MIN, PSIGN, WSIGN, ZSIGN)

This subroutine solves the 0-1 single knapsack problem

$$\text{maximize } Z = P(1)\,X(1) + \ldots + P(N)\,X(N)$$

$$\text{subject to} \quad W(1)\,X(1) + \ldots + W(N)\,X(N) \le C,$$

$$X(J) = 0 \text{ or } 1 \text{ for } J=1, \ldots, N$$

The program implements the branch-and-bound algorithm described in Section 2.5.2, and derives from an earlier code presented in S. Martello, P. Toth, "Algorithm for the solution of the 0-1 single knapsack problem", *Computing*, 1978.

The input problem must satisfy the conditions

(1) $2 \le N \le JDIM - 1$;

(2) P(J), W(J), C positive integers;

(3) $\max(W(J)) \le C$;

(4) $W(1) + \ldots + W(N) > C$;

(5) $P(J)/W(J) \ge P(J+1)/W(J+1)$ for $J = 1, \ldots, N-1$.

MT1 calls 1 procedure: CHMT1.

The program is completely self-contained and communication to it is achieved solely through the parameter list of MT1.
No machine-dependent constant is used.

MT1 needs 8 arrays (P, W, X, XX, MIN, PSIGN, WSIGN and ZSIGN) of length at least N + 1.

Meaning of the input parameters:

N = number of items;

P(J) = profit of item J (J = 1, ..., N);

W(J) = weight of item J (J = 1, ..., N);

C = capacity of the knapsack;

JDIM = dimension of the 8 arrays;

JCK = 1 if check on the input data is desired,
= 0 otherwise.

Meaning of the output parameters:

Z = value of the optimal solution if Z > 0,
= error in the input data (when JCK = 1) if Z < 0:
condition −Z is violated;

X(J) = 1 if item J is in the optimal solution,
= 0 otherwise.

Arrays XX, MIN, PSIGN, WSIGN and ZSIGN are dummy.

All the parameters are integer. On return of MT1 all the input parameters are unchanged.

```
INTEGER P(JDIM), W(JDIM), X(JDIM), C, Z
INTEGER XX(JDIM), MIN(JDIM)
INTEGER PSIGN(JDIM), WSIGN(JDIM), ZSIGN(JDIM)
```

A.2.2 Code MT1R

```
SUBROUTINE MT1R (N, P, W, C, EPS, Z, X, JDIM, JCK,
                 XX, MIN, PSIGN, WSIGN, ZSIGN, CRC, CRP)
```

This subroutine solves the 0-1 single knapsack problem with real parameters

$$\text{maximize } Z = P(1)\,X(1) + \ldots + P(N)\,X(N)$$

$$\text{subject to } \quad W(1)\,X(1) + \ldots + W(N)\,X(N) \leq C,$$

$$X(J) = 0 \text{ or } 1 \text{ for } J = 1, \ldots, N.$$

The program implements the branch-and-bound algorithm described in Section 2.5.2, and is a modified version of subroutine MT1.

The input problem must satisfy the conditions

(1) $2 \leq N \leq JDIM - 1$;

(2) P(J), W(J), C positive reals;

(3) max (W(J)) $\leq$ C;

(4) W(1) + ... + W(N) > C;

(5) P(J)/W(J) $\geq$ P(J + 1)/W(J + 1) for J = 1,..., N − 1.

MT1R calls 1 procedure: CHMT1R.

The program is completely self-contained and communication to it is achieved solely through the parameter list of MT1R.
No machine-dependent constant is used.

MT1R needs 10 arrays (P, W, X, XX, MIN, PSIGN, WSIGN, ZSIGN, CRC and CRP) of length at least N + 1.

Meaning of the input parameters:

N = number of items;

P(J) = profit of item J (J = 1,..., N);

W(J) = weight of item J (J = 1,..., N);

C = capacity of the knapsack;

EPS = tolerance (two positive values Q and R are considered equal if ABS(Q − R)/max (Q, R) $\leq$ EPS);

JDIM = dimension of the 10 arrays;

JCK = 1 if check on the input data is desired,
= 0 otherwise.

Meaning of the output parameters:

Z = value of the optimal solution if Z > 0,
= error in the input data (when JCK = 1) if Z < 0: condition −Z is violated;

X(J) = 1 if item J is in the optimal solution,
= 0 otherwise.

Arrays XX, MIN, PSIGN, WSIGN, ZSIGN, CRC and CRP are dummy.

Parameters N, X, JDIM, JCK, XX and ZSIGN are integer. Parameters P, W, C, Z,

MIN, PSIGN, WSIGN, CRC, CRP and EPS are real. On return of MT1R all the input parameters are unchanged.

```
REAL P(JDIM), W(JDIM)
INTEGER X(JDIM)
INTEGER XX(JDIM), ZSIGN(JDIM)
REAL MIN(JDIM), PSIGN(JDIM), WSIGN(JDIM), CRC(JDIM), CRP(JDIM)
```

A.2.3 Code MT2

SUBROUTINE MT2 (N, P, W, C, Z, X, JDIM, JFO, JFS, JCK, JUB, IA1, IA2, IA3, IA4, RA)

This subroutine solves the 0-1 single knapsack problem

$$\begin{aligned} \text{maximize } Z = \; & P(1)\,X(1) + \ldots + P(N)\,X(N) \\ \text{subject to } \; & W(1)\,X(1) + \ldots + W(N)\,X(N) \leq C, \\ & X(J) = 0 \text{ or } 1 \text{ for } J = 1, \ldots, N. \end{aligned}$$

The program implements the enumerative algorithm described in Section 2.9.3.

The input problem must satisfy the conditions

(1) $2 \leq N \leq JDIM - 3$;

(2) P(J), W(J), C positive integers;

(3) $\max(W(J)) \leq C$;

(4) $W(1) + \ldots + W(N) > C$;

and, if JFS = 1,

(5) $P(J)/W(J) \geq P(J+1)/W(J+1)$ for $J = 1, \ldots, N-1$.

MT2 calls 9 procedures: CHMT2, CORE, CORES, FMED, KP01M, NEWB, REDNS, REDS and SORTR.

The program is completely self-contained and communication to it is achieved solely through the parameter list of MT2.
No machine-dependent constant is used.

MT2 needs 8 arrays (P, W, X, IA1, IA2, IA3, IA4 and RA) of length at least N + 3.

Meaning of the input parameters:

N = number of items;

P(J) = profit of item J (J = 1,..., N);

W(J) = weight of item J (J = 1,..., N);

C = capacity of the knapsack;

JDIM = dimension of the 8 arrays;

JFO = 1 if optimal solution is required,
= 0 if approximate solution is required;

JFS = 1 if the items are already sorted according to decreasing profit per unit weight,
= 0 otherwise;

JCK = 1 if check on the input data is desired,
= 0 otherwise.

Meaning of the output parameters:

Z = value of the solution found if Z > 0,
= error in the input data (when JCK = 1) if Z < 0: condition −Z is violated;

X(J) = 1 if item J is in the solution found,
= 0 otherwise;

JUB = upper bound on the optimal solution value (to evaluate Z when JFO = 0).

Arrays IA1, IA2, IA3, IA4 and RA are dummy.

All the parameters but RA are integer. On return of MT2 all the input parameters are unchanged.

```
INTEGER P(JDIM), W(JDIM), X(JDIM), C, Z
DIMENSION IA1(JDIM), IA2(JDIM), IA3(JDIM), IA4(JDIM)
DIMENSION RA(JDIM)
```

A.3 BOUNDED AND UNBOUNDED KNAPSACK PROBLEM

A.3.1 Code MTB2

```
SUBROUTINE MTB2  (N, P, W, B, C, Z, X,
                  JDIM1, JDIM2, JFO, JFS, JCK, JUB,
                  ID1, ID2, ID3, ID4, ID5, ID6, ID7, RD8)
```

This subroutine solves the bounded single knapsack problem

$$\text{maximize } Z = P(1)\,X(1) + \ldots + P(N)\,X(N)$$

$$\text{subject to } W(1)\,X(1) + \ldots + W(N)\,X(N) \le C,$$

$$0 \le X(J) \le B(J) \quad \text{for } J = 1, \ldots, N,$$

$$X(J) \text{ integer} \quad \text{for } J = 1, \ldots, N.$$

The program implements the transformation method described in Section 3.2.

The problem is transformed into an equivalent 0-1 knapsack problem and then solved through subroutine MT2. The user must link MT2 and its subroutines to this program.

The input problem must satisfy the conditions

(1) $2 \le N \le JDIM1 - 1$;

(2) P(J), W(J), B(J), C positive integers;

(3) $\max (B(J)W(J)) \le C$;

(4) $B(1)W(1) + \ldots + B(N)W(N) > C$;

(5) $2 \le N + (LOG2(B(1)) + \ldots + LOG2(B(N))) \le JDIM2 - 3$;

and, if JFS = 1,

(6) $P(J)/W(J) \ge P(J + 1)/W(J + 1)$ for $J = 1, \ldots, N - 1$.

MTB2 calls 4 procedures: CHMTB2, SOL, TRANS and MT2 (external).

Communication to the program is achieved solely through the parameter list of MTB2.
No machine-dependent constant is used.

MTB2 needs

4 arrays (P, W, B and X) of length at least JDIM1;

8 arrays (ID1, ID2, ID3, ID4, ID5, ID6, ID7 and RD8) of length at least JDIM2.

Meaning of the input parameters:

N = number of item types;

P(J) = profit of each item of type J (J = 1, ..., N);

W(J) = weight of each item of type J (J = 1, ..., N);

B(J) = number of items of type J available (J = 1, ..., N);

C = capacity of the knapsack;

JDIM1 = dimension of arrays P, W, B, X;

JDIM2 = dimension of arrays ID1, ID2, ID3, ID4, ID5, ID6, ID7, RD8;

JFO = 1 if optimal solution is required,
= 0 if approximate solution is required;

JFS = 1 if the items are already sorted according to decreasing profit per unit weight (suggested for large B(J) values),
= 0 otherwise;

JCK = 1 if check on the input data is desired,
= 0 otherwise.

Meaning of the output parameters:

Z = value of the solution found if Z > 0,
= error in the input data (when JCK = 1) if Z < 0: condition −Z is violated;

X(J) = number of items of type J in the solution found;

JUB = upper bound on the optimal solution value (to evaluate Z when JFO = 0).

Arrays ID1, ID2, ID3, ID4, ID5, ID6, ID7 and RD8 are dummy.

All the parameters but RD8 are integer. On return of MTB2 all the input parameters are unchanged.

```
INTEGER P(JDIM1), W(JDIM1), B(JDIM1), X(JDIM1), C, Z
INTEGER ID1(JDIM2), ID2(JDIM2), ID3(JDIM2), ID4(JDIM2)
INTEGER ID5(JDIM2), ID6(JDIM2), ID7(JDIM2)
REAL RD8(JDIM2)
```

A.3.2 Code MTU2

```
SUBROUTINE MTU2 (N, P, W, C, Z, X,
                 JDIM, JFO, JCK, JUB,
                 PO, WO, XO, RR, PP)
```

This subroutine solves the unbounded single knapsack problem

$$\text{maximize } Z = P(1)\,X(1) + \ldots + P(N)\,X(N)$$

$$\text{subject to } W(1)\,X(1) + \ldots + W(N)\,X(N) \leq C,$$

$$X(J) \geq 0 \text{ and integer for } J = 1, \ldots, N.$$

The program implements the enumerative algorithm described in Section 3.6.3.

The input problem must satisfy the conditions

(1) $2 \leq N \leq JDIM - 1$;

(2) P(J), W(J), C positive integers;

(3) max (W(J)) $\leq$ C.

MTU2 calls 5 procedures: CHMTU2, KSMALL, MTU1, REDU and SORTR.
KSMALL calls 8 procedures: BLD, BLDF, BLDS1, DETNS1, DETNS2, FORWD, MPSORT and SORT7.

The program is completely self-contained and communication to it is achieved solely through the parameter list of MTU2.
No machine-dependent constant is used.

MTU2 needs 8 arrays (P, W, X, PO, WO, XO, RR and PP) of length at least JDIM.

Meaning of the input parameters:

N = number of item types;

P(J) = profit of each item of type J (J = 1,..., N);

W(J) = weight of each item of type J (J = 1,..., N);

C = capacity of the knapsack;

JDIM = dimension of the 8 arrays;

JFO = 1 if optimal solution is required,
= 0 if approximate solution is required;

JCK = 1 if check on the input data is desired,
= 0 otherwise.

Meaning of the output parameters:

Z = value of the solution found if $Z > 0$,
= error in the input data (when JCK = 1) if $Z < 0$:
condition $-Z$ is violated;

X(J) = number of items of type J in the solution found;

JUB = upper bound on the optimal solution value
(to evaluate Z when JFO = 0).

Arrays PO, WO, XO, RR and PP are dummy.

All the parameters but XO and RR are integer. On return of MTU2 all the input parameters are unchanged.

```
INTEGER P(JDIM), W(JDIM), X(JDIM)
INTEGER PO(JDIM), WO(JDIM), PP(JDIM), C, Z
REAL RR(JDIM), XO(JDIM)
```

A.4 SUBSET-SUM PROBLEM

A.4.1 Code MTSL

```
SUBROUTINE MTSL (N, W, C, Z, X, JDN, JDD, ITMM, JCK,
                 WO, IND, XX, WS, ZS, SUM,
                 TD1, TD2, TD3)
```

This subroutine solves the subset-sum problem

$$\text{maximize } Z = W(1)\,X(1) + \ldots + W(N)\,X(N)$$

$$\text{subject to} \quad W(1)\,X(1) + \ldots + W(N)\,X(N) \leq C,$$
$$X(J) = 0 \text{ or } 1 \qquad \text{for } J = 1, \ldots, N.$$

The program implements the mixed algorithm described in Section 4.2.3.

The input problem must satisfy the conditions

(1) $2 \leq N \leq JDN - 1$;

(2) W(J), C positive integers;

(3) $\max(W(J)) < C$;

(4) $W(1) + \ldots + W(N) > C$.

MTSL calls 8 procedures: CHMTSL, DINSM, MTS, PRESP, SORTI, TAB, UPSTAR and USEDIN.

If not present in the library of the host, the user must supply an integer function JIAND(I1, I2) which sets JIAND to the bit-by-bit logical AND of I1 and I2.

Communication to the program is achieved solely through the parameter list of MTSL.
No machine-dependent constant is used.

MTSL needs

2 arrays (W and X) of length at least JDN;
6 arrays (WO, IND, XX, WS, ZS and SUM) of length at least ITMM;
3 arrays (TD1, TD2 and TD3) of length at least JDD × 2.

Meaning of the input parameters:

N = number of items;

W(J) = weight of item J (J = 1, ..., N);

C = capacity;

JDN = dimension of arrays W and X;

JDD = maximum length of the dynamic programming lists (suggested value JDD = 5000);

ITMM = (maximum number of items in the core problem) + 1; ITMM = JDN in order to be sure that the optimal solution is found. ITMM < JDN (suggested value ITMM = 91) produces an approximate solution which is almost always optimal (to check optimality, see whether Z = C);

JCK = 1 if check on the input data is desired,
= 0 otherwise.

Meaning of the output parameters:

Z = value of the solution found if Z > 0,
= error in the input data (when JCK = 1) if Z < 0: condition −Z is violated;

X(J) = 1 if item J is in the solution found,
= 0 otherwise.

Meaning of the internal variables which could be altered by the user:

IT = length of the initial core problem (suggested value IT = 30);

ID = increment of the length of the core problem (suggested value ID = 30);

M2 = number of items to be used for the second dynamic programming list; it must be $2 \leq M2 \leq \min(31, N - 4)$ (suggested value M2 = min (2.5 ALOG10 (max (W(J))), 0.8 N)). M1, the number of items to be used for the first dynamic programming list, is automatically determined;

PERS = value used to determine $\overline{c}$ according to the formula given in Section 4.2.2 (suggested value PERS = 1.3).

Arrays WO, IND, XX, WS, ZS, SUM, TD1, TD2 and TD3 are dummy.

All the parameters are integer. On return of MTSL all the input parameters are unchanged.

```
INTEGER W(JDN), X(JDN), C, Z
INTEGER WO(ITMM), IND(ITMM), XX(ITMM)
INTEGER WS(ITMM), ZS(ITMM), SUM(ITMM)
INTEGER TD1(JDD,2), TD2(JDD,2), TD3(JDD,2)
```

A.5 BOUNDED AND UNBOUNDED CHANGE-MAKING PROBLEM

A.5.1 Code MTC2

```
SUBROUTINE MTC2  (N, W, C, Z, X, JDN, JDL, JFO, JCK,
                  XX, WR, PR, M, L)
```

This subroutine solves the unbounded change-making problem

$$\text{minimize } Z = X(1) + \ldots + X(N)$$

$$\text{subject to} \quad W(1)\,X(1) + \ldots + W(N)\,X(N) = C,$$

$$X(J) \geq 0 \text{ and integer for } J = 1, \ldots, N.$$

The program implements the enumerative algorithm described in Section 5.6.

The input problem must satisfy the conditions

(1) $2 \leq N \leq JDN - 1$;

(2) W(J), C positive integers;

(3) max (W(J)) $<$ C.

MTC2 calls 5 procedures: CHMTC2, COREC, MAXT, MTC1 and SORTI.

The program is completely self-contained and communication to it is achieved solely through the parameter list of MTC2.
No machine-dependent constant is used.

MTC2 needs

5 arrays (W, X, XX, WR and PR) of length at least JDN;
2 arrays (M and L) of length at least JDL.

Meaning of the input parameters:

N = number of item types;

W(J) = weight of each item of type J (J = 1,..., N);

C = capacity;

JDN = dimension of arrays W, X, XX, WR and PR;

JDL = dimension of arrays M and L (suggested value JDL = max (W(J)) − 1; if the core memory is not enough, JDL should be set to the largest possible value);

JFO = 1 if optimal solution is required,
= 0 if approximate solution is required
(at most 100 000 backtrackings are performed);

JCK = 1 if check on the input data is desired,
= 0 otherwise.

Meaning of the output parameters:

Z = value of the solution found if Z > 0,
= no feasible solution exists if Z = 0,
= error in the input data (when JCK = 1) if Z < 0:
condition −Z is violated;

X(J) = number of items of type J in the solution found.

Arrays XX, M, L, WR and PR are dummy.

All the parameters are integer. On return of MTC2 all the input parameters are unchanged.

```
INTEGER W(JDN), X(JDN), C, Z
INTEGER XX(JDN), WR(JDN), PR(JDN)
INTEGER M(JDL), L(JDL)
```

A.5.2 Code MTCB

```
SUBROUTINE MTCB  (N, W, B, C, Z, X, JDN, JDL, JFO, JCK,
                  XX, WR, BR, PR, M, L)
```

This subroutine solves the bounded change-making problem

$$\text{minimize } Z = X(1) + \ldots + X(N)$$

$$\text{subject to} \quad W(1)\,X(1) + \ldots + W(N)\,X(N) = C,$$

$$0 \le X(J) \le B(J) \quad \text{for } J = 1, \ldots, N,$$

$$X(J) \text{ integer} \quad \text{for } J = 1, \ldots, N.$$

The program implements the branch-and-bound algorithm described in Section 5.8.

The input problem must satisfy the conditions

(1) $2 \le N \le JDN - 1$;

(2) W(J), B(J), C positive integers;

(3) max (W(J)) < C;

(4) $B(J)\ W(J) \leq C$ for $J = 1, \ldots, N$;

(5) $B(1)\ W(1) + \ldots + B(N)\ W(N) > C$.

MTCB calls 3 procedures: CHMTCB, CMPB and SORTI.

The program is completely self-contained and communication to it is achieved solely through the parameter list of MTCB.
No machine-dependent constant is used.

MTCB needs

7 arrays (W, B, X, XX, WR, BR and PR) of length at least JDN;
2 arrays (M and L) of length at least JDL.

Meaning of the input parameters:

N = number of item types;

W(J) = weight of each item of type J ($J = 1, \ldots, N$);

B(J) = number of available items of type J ($J = 1, \ldots, N$);

C = capacity;

JDN = dimension of arrays W, B, X, XX, WR, BR and PR;

JDL = dimension of arrays M and L (suggested value JDL = max (W(J)) − 1; if the core memory is not enough, JDL should be set to the largest possible value);

JFO = 1 if optimal solution is required,
= 0 if approximate solution is required
(at most 100 000 backtrackings are performed);

JCK = 1 if check on the input data is desired,
= 0 otherwise.

Meaning of the output parameters:

Z = value of the solution found if $Z > 0$,
= no feasible solution exists if $Z = 0$,
= error in the input data (when JCK = 1) if $Z < 0$:
condition $-Z$ is violated;

X(J) = number of items of type J in the solution found.

Arrays XX, M, L, WR, BR and PR are dummy.

All the parameters are integer. On return of MTCB all the input parameters are unchanged.

```
INTEGER W(JDN), B(JDN), X(JDN), C, Z
INTEGER XX(JDN), WR(JDN), BR(JDN), PR(JDN)
INTEGER M(JDL), L(JDL)
```

A.6 0-1 MULTIPLE KNAPSACK PROBLEM

A.6.1 Code MTM

SUBROUTINE MTM (N, M, P, W, C, Z, X, BACK, JCK, JUB)

This subroutine solves the 0-1 multiple knapsack problem

```
maximize Z =  P(1) (Y(1, 1)  + ... + Y(M, 1)) +
                                ...           +
              P(N) (Y(1, N)  + ... + Y(M, N))

subject to    W(1) Y(I, 1) + ... + W(N) Y(I, N) <= C(I)
                                          for I = 1,..., M,

              Y(1, J) + ... + Y(M, J) <= 1    for J = 1,..., N,

              Y(I, J) = 0 or 1  for I = 1,..., M, J = 1,..., N.
```

The program implements the enumerative algorithm described in Section 6.4.3, and derives from an earlier code presented in S. Martello, P. Toth, "Algorithm 632. A program for the 0-1 multiple knapsack problem", *ACM Transactions on Mathematical Software*, 1985.

The input problem must satisfy the conditions

(1) 2 ≤ N ≤ MAXN and 1 ≤ M ≤ MAXM, where MAXN and MAXM are defined by the first two executable statements;

(2) P(J), W(J) and C(I) positive integers;

(3) min (C(I)) ≥ min (W(J));

(4) max (W(J)) ≤ max (C(I));

(5) max (C(I)) < W(1) + ... + W(N);

(6) P(J)/W(J) ≥ P(J + 1)/W(J + 1) for J = 1,..., N − 1;

(7) C(I) ≤ C(I + 1) for I = 1,..., M − 1.

MTM calls 5 procedures: CHMTM, PAR, PI, SIGMA and SKP.

The program is completely self-contained and communication to it is achieved solely through the parameter list of MTM.
No machine-dependent constant is used.

MTM needs

5 arrays (C, F, PBL, Q and V) of length at least M;
8 arrays (P, W, X, UBB, BS, XS, LX and LXI) of length at least N;
3 arrays (B, PS and WS) of length at least N + 1;
3 arrays (BB, XC and XL) of length at least M × N;
1 array (BL) of length at least M × (N + 1);
5 arrays (D, MIN, PBAR, WBAR and ZBAR) of length at least N (for internal use in subroutine SKP).

The arrays are currently dimensioned to allow problems for which M ≤ 10 and N ≤ 1000. Changing such dimensions also requires changing the dimension of BS, PS, WS, XS, LX and LXI in subroutine SIGMA, of BB, BL, XL, BS, PS, WS and XS in subroutine PI, of BB, LX and LXI in subroutine PAR, of D, MIN, PBAR, WBAR and ZBAR in subroutine SKP. In addition, the values of MAXN and MAXM must be conveniently defined.

Meaning of the input parameters:

N = number of items;

M = number of knapsacks;

P(J) = profit of item J (J = 1,..., N);

W(J) = weight of item J (J = 1,..., N);

C(I) = capacity of knapsack I (I = 1,..., M);

BACK = −1 if exact solution is required,
= maximum number of backtrackings to be performed, if heuristic solution is required;

JCK = 1 if check on the input data is desired,
= 0 otherwise.

Meaning of the output parameters:

Z = value of the solution found if Z > 0,
= error in the input data (when JCK = 1) if Z < 0: condition −Z is violated;

X(J) = 0 if item J is not in the solution found (Y(I, J) = 0 for all I),
= knapsack where item J is inserted, otherwise (Y(X(J), J) = 1);

JUB = upper bound on the optimal solution value (to evaluate Z when BACK ≥ 0 on input).

All the parameters are integer. On return of MTM all the input parameters are unchanged except BACK (= number of backtrackings performed).

```
INTEGER P(1000), W(1000), C(10), X(1000), Z, BACK
INTEGER BB(10,1000), BL(10,1001), XC(10,1000), XL(10,1000)
INTEGER B(1001), UBB(1000), F(10), PBL(10), Q(10), V(10)
INTEGER BS, PS, WS, XS
COMMON /SNGL/ BS(1000), PS(1001), WS(1001), XS(1000)
COMMON /PUB/ LX(1000), LXI(1000), LR, LRI, LUBI
```

A.6.2 Code MTHM

```
SUBROUTINE MTHM (N, M, P, W, C, Z, X, JDN, JDM, LI, JCK,
                 CR, MIN, XX, X1, F)
```

This subroutine heuristically solves the 0-1 multiple knapsack problem

```
maximize Z = P(1) (Y(1, 1)  + ... + Y(M, 1)) +
                              ...             +
             P(N) (Y(1, N) + ... + Y(M, N))

subject to   W(1) Y(I, 1) + ... + W(N) Y(I, N) ≤ C(I)
                                        for I = 1,..., M,

             Y(1,J) + ... + Y(M,J) ≤ 1   for J = 1,..., N,

             Y(I,J) = 0 or 1  for I = 1,..., M,  J = 1,..., N.
```

The program implements the polynomial-time algorithms described in Section 6.6.2, and derives from an earlier code presented in S. Martello, P. Toth, "Heuristic algorithms for the multiple knapsack problem", *Computing*, 1981.

The input problem must satisfy the conditions

(1) 2 ≤ N ≤ JDN − 1 and 1 ≤ M ≤ JDM − 1;

(2) P(J), W(J) and C(I) positive integers;

(3) min (C(I)) ≥ min (W(J));

(4) max (W(J)) ≤ max (C(I));

(5) max (C(I)) < W(1) + ... + W(N);

(6) P(J)/W(J) ≥ P(J+1)/W(J+1) for J = 1,..., N − 1;

(7) C(I) ≤ C(I+1) for I = 1,..., M − 1.

MTHM can call 6 subroutines:

CHMTHM to check the input data;
MGR1 or MGR2 to find an initial feasible solution;
REARR to re-arrange a feasible solution;
IMPR1 and IMPR2 to improve on a feasible solution.

The user selects the sequence of calls through input parameters.

The program is completely self-contained and communication to it is achieved solely through the parameter list of MTHM.
The only machine-dependent constant is used to define INF (first executable statement), which must be set to a large positive integer value.

MTHM needs

6 arrays (P, W, X, MIN, XX and X1) of length at least JDN;
2 arrays (C and CR) of length at least JDM;
1 array (F) of length at least JDM $\times$ JDM.

In addition, subroutine MGR2 uses

7 arrays of length 5;
1 array of length 201;
1 array of length 5 $\times$ 200.

Subroutine MGR2 is called only when M $\leq$ 5 and N $\leq$ 200.

Meaning of the input parameters:

N = number of items;

M = number of knapsacks;

P(J) = profit of item J (J = 1,..., N);

W(J) = weight of item J (J = 1,..., N);

C(I) = capacity of knapsack I (I = 1,..., M);

JDN = dimension of arrays P, W, X, MIN, XX and X1;

JDM = dimension of arrays C, CR and F;

LI = 0 to output the initial feasible solution,
= 1 to also perform subroutines REARR and IMPR1,
= 2 to also perform subroutines REARR, IMPR1 and IMPR2;

JCK = 1 if check on the input data is desired,
= 0 otherwise.

Meaning of the output parameters:

Z = value of the solution found if Z > 0,
= error in the input data (when JCK = 1) if Z < 0:
condition −Z is violated;

X(J) = 0 if item J is not in the solution found
(i.e. if Y(I, J) = 0 for all I),
= knapsack where item J is inserted, otherwise
(i.e. if Y(X(J), J) = 1).

Arrays CR, MIN, XX, X1 and F are dummy.

All the parameters are integer. On return of MTHM all the input parameters are unchanged.

```
INTEGER P(JDN), W(JDN), X(JDN), C(JDM), Z
INTEGER MIN(JDN), XX(JDN), X1(JDN), CR(JDM)
INTEGER F(JDM, JDM)
```

A.7 GENERALIZED ASSIGNMENT PROBLEM

A.7.1 Code MTG

```
SUBROUTINE MTG (N, M, P, W, C, MINMAX,
                Z, XSTAR, BACK, JCK, JB)
```

This subroutine solves the generalized assignment problem

$$\begin{aligned} \text{opt } Z = \; & P(1, 1)\, X(1, 1) + \ldots + P(1, N)\, X(1, N) \; + \\ & \ldots \; + \\ & P(M, 1)\, X(M, 1) + \ldots + P(M, N)\, X(M, N) \end{aligned}$$

(where opt = min if MINMAX = 1, opt = max if MINMAX = 2)

$$\text{subject to } \; W(I, 1)\, X(I, 1) + \ldots + W(I, N)\, X(I, N) \le C(I) \quad \text{for } I = 1, \ldots, M,$$

$$X(1, J) + \ldots + X(M, J) = 1 \quad \text{for } J = 1, \ldots, N,$$

$$X(I, J) = 0 \text{ or } 1 \quad \text{for } I = 1, \ldots, M, \; J = 1, \ldots, N.$$

The program implements the branch-and-bound algorithm described in Sections 7.3–7.5.

The input problem must satisfy the conditions

(1) $2 \le M \le$ JDIMR;

(2) $2 \le N \le$ JDIMC (JDIMR and JDIMC are defined by the first two executable statements);

(3) $M \le$ JDIMPC (JDIMPC, defined by the third executable statement, is used for packing array Y, and cannot be greater than (number of bits of the host) -2; if

a higher value is desired, subroutines YDEF and YUSE must be re-structured accordingly);

(4) P(I, J), W(I, J) and C(I) positive integers;

(5) W(I, J) $\leq$ C(I) for at least one I, for J = 1,..., N;

(6) C(I) $\geq$ min (W(I, J)) for I = 1,..., M.

In addition, it is required that

(7) (maximum level of the decision-tree) $\leq$ JNLEV. (JNLEV is defined by the fourth executable statement.)

MTG calls 24 procedures: CHMTG, DEFPCK, DMIND, FEAS, GHA, GHBCD, GHX, GR1, GR2, HEUR, KPMAX, KPMIN, PEN0, PEN1, PREPEN, SKP, SORTI, SORTR, TERMIN, TRIN, UBFJV, UBRS, YDEF and YUSE.

If not present in the library of the host, the user must supply an integer function JIAND(I1, I2) which sets JIAND to the bit-by-bit logical AND of I1 and I2. Such function is used in subroutines YDEF and YUSE.

Communication to the program is achieved solely through the parameter list of MTG.
No machine-dependent constant is used.

MTG needs

17 arrays (C, DD, UD, Q, PACKL, IP, IR, IL, IF, WOBBL, KQ, FLREP, DMYR1, DMYR2, DMYR3, DMYR4 and DMYR5) of length at least M;
25 arrays (XSTAR, XS, BS, B, KA, XXS, IOBBL, JOBBL, BEST, XJJUB, DS, DMYC1, DMYC2, DMYC3, DMYC4, DMYC5, DMYC6, DMYC7, DMYC8, DMYC9, DMYC10, DMYC11, DMYC12, DMYC13 and DMYCR1) of length at least N;
4 arrays (PS, WS, DMYCC1 and DMYCC2) of length at least N + 1;
6 arrays (E, CC, CS, TYPE, US and UBL) of length at least JNLEV;
7 arrays (P, W, A, X, PAK, KAP and MIND) of length at least M $\times$ N;
5 arrays (D, VS, V, LB and UB) of length at least JNLEV $\times$ M;
1 array (Y) of length at least JNLEV $\times$ N;
2 arrays (MASK1 and ITWO) of length at least JDIMPC.

The arrays are currently dimensioned to allow problems for which

M $\leq$ 10,
N $\leq$ 100,
JNLEV $\leq$ 150,

on a 32-bit computer (so, in the calling program, arrays P and W must be dimensioned at (10,100)). Changing such limits necessitates changing the dimension of all the arrays in subroutine MTG and in COMMON /PACK/ (which is included in subroutines MTG, YDEF and YUSE), as well as the four first executable statements.

Meaning of the input parameters:

N = number of items;

M = number of knapsacks;

P(I, J) = profit of item J if assigned to knapsack I
(I = 1, . . . , M; J = 1, . . . , N);

W(I, J) = weight of item J if assigned to knapsack I
(I = 1, . . . , M; J = 1, . . . , N);

C(I) = capacity of knapsack I (I = 1, . . . , M);

MINMAX = 1 if the objective function must be minimized,
= 2 if the objective function must be maximized;

BACK = –1 if exact solution is required,
= maximum number of backtrackings to be performed, if heuristic solution is required;

JCK = 1 if check on the input data is desired,
= 0 otherwise.

Meaning of the output parameters:

Z = value of the solution found if Z $>$ 0,
= 0 if no feasible solution exists,
= error in the input data (when JCK = 1) if Z $<$ 0: condition $-$Z is violated;

XSTAR(J) = knapsack where item J is inserted in the solution found;

JB = lower bound (if MINMAX = 1) or upper bound (if MINMAX = 2) on the optimal solution value (to evaluate Z when BACK $\geq$ 0 on input).

All the parameters are integer. On return of MTG all the input parameters are unchanged, with the following two exceptions. BACK gives the number of backtrackings performed; P(I, J) is set to 0 for all pairs (I, J) such that W(I, J) $>$ C(I).

```
INTEGER P(10,100), W(10,100), C(10), XSTAR(100), Z, BACK
INTEGER DD(10), UD(10), Q(10), PAKL(10), IP(10), IR(10)
```

```
INTEGER IL(10), IF(10), WOBBL(10), KQ(10), FLREP(10)
INTEGER XS(100), BS(100), B(100), KA(100), XXS(100)
INTEGER IOBBL(100), JOBBL(100), BEST(100), XJJUB(100)
REAL DS(100)
INTEGER PS(101), WS(101)
INTEGER E(150), CC(150), CS(150)
INTEGER TYPE(150), US(150), UBL(150)
INTEGER A(10,100), X(10,100)
INTEGER PAK(10,100), KAP(10,100), MIND(10,100)
INTEGER D(150,10), VS(150,10)
INTEGER V(150,10), LB(150,10), UB(150,10)
INTEGER Y
INTEGER DMYR1(10), DMYR2(10), DMYR3(10)
INTEGER DMYR4(10), DMYR5(10)
INTEGER DMYC1(100), DMYC2(100), DMYC3(100)
INTEGER DMYC4(100), DMYC5(100), DMYC6(100)
INTEGER DMYC7(100), DMYC8(100), DMYC9(100)
INTEGER DMYC10(100), DMYC11(100), DMYC12(100)
INTEGER DMYC13(100)
INTEGER DMYCC1(101), DMYCC2(101)
REAL DMYCR1(100)
COMMON /PACK/ MASK1(30), ITWO(30), MASK, Y(150,100)
```

A.7.2 Code MTHG

```
SUBROUTINE MTHG (N, M, P, W, C, MINMAX,
                 Z, XSTAR, JCK)
```

This subroutine heuristically solves the generalized assignment problem

$$\text{opt } Z = P(1, 1)\, X(1, 1) + \ldots + P(1, N)\, X(1, N) + \ldots + P(M, 1)\, X(M, 1) + \ldots + P(M, N)\, X(M, N)$$

(where opt = min if MINMAX = 1, opt = max if MINMAX = 2)

$$\text{subject to } W(I, 1)\, X(I, 1) + \ldots + W(I, N)\, X(I, N) \leq C(I) \quad \text{for } I = 1, \ldots, M,$$

$$X(1, J) + \ldots + X(M, J) = 1 \quad \text{for } J = 1, \ldots, N,$$

$$X(I, J) = 0 \text{ or } 1 \text{ for } I = 1, \ldots, M, \; J = 1, \ldots, N.$$

The program implements the polynomial-time algorithms described in Section 7.4.

The input problem must satisfy the conditions

(1) $2 \leq M \leq$ JDIMR;

(2) $2 \leq N \leq$ JDIMC (JDIMR and JDIMC are defined by the first two executable statements);

(3) P(I, J), W(I, J) and C(I) positive integers;

(4) W(I, J) $\leq$ C(I) for at least one I, for J = 1, ..., N;

(5) C(I) $\geq$ min (W(I, J)) for I = 1, ..., M.

MTHG calls 6 procedures: CHMTHG, FEAS, GHA, GHBCD, GHX and TRIN.

Communication to the program is achieved solely through the parameter list of MTHG.
No machine-dependent constant is used.

MTHG needs

6 arrays (C, DMYR1, DMYR2, DMYR3, DMYR4 and DMYR5) of length at least JDIMR;
7 arrays (XSTAR, BEST, DMYC1, DMYC2, DMYC3, DMYC4 and DMYCR1) of length at least JDIMC;
3 arrays (P, W and A) of length at least JDMR × JDIMC.

The arrays are currently dimensioned to allow problems for which

$M \leq 50$,
$N \leq 500$

(so, in the calling program, arrays P and W must be dimensioned at (50,500)). Changing such limits necessitates changing the dimension of all the arrays in subroutine MTHG, as well as the first two executable statements.

Meaning of the input parameters:

N = number of items;

M = number of knapsacks;

P(I, J) = profit of item J if assigned to knapsack I (I = 1, ..., M; J = 1, ..., N);

W(I, J) = weight of item J if assigned to knapsack I (I = 1, ..., M; J = 1, ..., N);

C(I) = capacity of knapsack I (I = 1, ..., M);

MINMAX = 1 if the objective function must be minimized,
= 2 if the objective function must be maximized;

JCK = 1 if check on the input data is desired,
= 0 otherwise.

Meaning of the output parameters:

Z = value of the solution found if Z > 0,
= 0 if no feasible solution is found,
= error in the input data (when JCK = 1) if Z < 0: condition −Z is violated;

XSTAR(J) = knapsack where item J is inserted in the solution found.

All the parameters are integer. On return of MTHG all the input parameters are unchanged, but P(I, J) is set to 0 for all pairs (I, J) such that W(I, J) > C(I).

```
INTEGER P(50,500), W(50,500), C(50), XSTAR(500), Z
INTEGER BEST(500)
INTEGER A(50,500)
INTEGER DMYR1(50), DMYR2(50), DMYR3(50)
INTEGER DMYR4(50), DMYR5(50)
INTEGER DMYC1(500), DMYC2(500), DMYC3(500)
INTEGER DMYC4(500)
REAL DMYCR1(500)
```

A.8 BIN-PACKING PROBLEM

A.8.1 Code MTP

```
SUBROUTINE MTP  (N, W, C, Z, XSTAR,
                 JDIM, BACK, JCK, LB,
                 WR, XSTARR, DUM, RES, REL, X, R, WA,
                 WB, KFIX, FIXIT, XRED, LS, LSB, XHEU)
```

This subroutine solves the bin packing problem

minimize Z = Y(1) + ... + Y(N)

subject to W(1) X(I, 1) + ... + W(N) X(I, N) ≤ C Y(I) for I = 1,..., N,

X(1, J) + ... + X(M, J) = 1 for J = 1,..., N,

Y(I) = 0 or 1 for I = 1,..., N,

X(I, J) = 0 or 1 for I = 1,..., N, J = 1,..., N

(i.e., minimize the number of bins of capacity C needed to allocate N items of size W(1),..., W(N)).

The program implements the branch-and-bound algorithm described in Section 8.5.

The input problem must satisfy the conditions

(1) $2 \leq N \leq JDIM$;

(2) W(J) and C positive integers;

(3) $W(J) \leq C$ for $J = 1, \ldots, N$;

(4) $W(J) \geq W(J+1)$ for $J = 1, \ldots, N-1$.

In the output solution (see below) the Z lowest indexed bins are used.

MTP calls 14 procedures: CHMTP, ENUMER, FFDLS, FIXRED, HBFDS, INSERT, LCL2, L2, L3, MWFDS, RESTOR, SEARCH, SORTI2 and UPDATE.

Communication to the program is achieved solely through the parameter list of MTP.
No machine-dependent constant is used.

MTP needs

17 arrays (W, XSTAR, WR, XSTARR, DUM, RES, REL, X, R, WA, WB, KFIX, FIXIT, XRED, LS, LSB and XHEU) of length at least JDIM.

Meaning of the input parameters:

N = number of items;

W(J) = weight of item J;

C = capacity of the bins;

JDIM = dimension of the 17 arrays;

BACK = −1 if exact solution is required,
= maximum number of backtrackings to be performed, if heuristic solution is required;

JCK = 1 if check on the input data is desired,
= 0 otherwise.

Meaning of the output parameters:

Z = value of the solution found if $Z > 0$,
= error in the input data (when JCK = 1) if $Z < 0$:
condition $-Z$ is violated;

XSTAR(J) = bin where item J is inserted in the solution found;

LB = lower bound on the optimal solution value
(to evaluate Z when $BACK \geq 0$ on input).

All the arrays except W and XSTAR are dummy.

All the parameters are integer. On return of MTP all the input parameters are unchanged except BACK, which gives the number of backtrackings performed.

```
INTEGER W(JDIM), XSTAR(JDIM), C, Z, BACK
INTEGER WR(JDIM), XSTARR(JDIM), DUM(JDIM)
INTEGER RES(JDIM), REL(JDIM), X(JDIM), R(JDIM)
INTEGER WA(JDIM), WB(JDIM), KFIX(JDIM)
INTEGER FIXIT(JDIM), XRED(JDIM), LS(JDIM)
INTEGER LSD(JDIM), XHEU(JDIM)
```

Glossary

$O(f(n))$	order of $f(n)$
$\|S\|$	cardinality of set S
$r(A)$	worst-case performance ratio of algorithm A
$\varepsilon(A)$	worst-case relative error of algorithm A
$\rho(B)$	worst-case performance ratio of bound B
$\lfloor a \rfloor$	largest integer not greater than a
$\lceil a \rceil$	smallest integer not less than a
$z(P)$	optimal solution value of problem P
$C(P)$	continuous relaxation of problem P
$L(P, \lambda)$	Lagrangian relaxation of problem P through multiplier λ
$S(P, \pi)$	surrogate relaxation of problem P through multiplier π
$i \pmod j$	$i - \lfloor i/j \rfloor j$ (i, j positive integers)
arg max $\{s_1, \ldots, s_n\}$	index k such that $s_k \geq s_i$ for $i = 1, \ldots, n$
max $\{s_1, \ldots, s_n\}$	$s_{\arg\max \{s_1, \ldots, s_n\}}$
arg $\max_2$ $\{s_1, \ldots, s_n\}$	arg max $(\{s_1, \ldots, s_n\} / \{s_{\arg\max \{s_1, \ldots, s_n\}}\})$
$\max_2 \{s_1, \ldots, s_n\}$	$s_{\arg\max_2 \{s_1, \ldots, s_n\}}$

arg min, min, arg $\min_2$, $\min_2$ are immediate extensions of the above

Bibliography

A.V. Aho, J.E. Hopcroft, J.D. Ullman (1983). *Data Structures and Algorithms,* Addison-Wesley, Reading, MA.

J.H. Ahrens, G. Finke (1975). Merging and sorting applied to the 0-1 knapsack problem. *Operations Research* **23**, 1099–1109.

L. Aittoniemi (1982). Computational comparison of knapsack algorithms, Presented at XIth International Symposium on Mathematical Programming, Bonn, August 23–27.

L. Aittoniemi, K. Oehlandt (1985). A note on the Martello–Toth algorithm for one-dimensional knapsack problems. *European Journal of Operational Research* **20**, 117.

R.D. Armstrong, D.S. Kung, P. Sinha, A.A. Zoltners (1983). A computational study of a multiple-choice knapsack algorithm. *ACM Transactions on Mathematical Software* **9**, 184–198.

G. d'Atri (1979). Analyse probabiliste du problème du sac-à-dos. *Thèse*, Université de Paris VI.

G. d'Atri, C. Puech (1982). Probabilistic analysis of the subset-sum problem. *Discrete Applied Mathematics* **4**, 329–334.

D. Avis (1980). Theorem 4. In V. Chvátal. Hard knapsack problems, *Operations Research* **28**, 1410–1411.

L.G. Babat (1975). Linear functions on the *N*-dimensional unit cube. *Doklady Akademiia Nauk SSSR* **222**, 761–762.

A. Bachem, M. Grötschel (1982). New aspects of polyhedral theory. In B. Korte (ed.), *Modern Applied Mathematics, Optimization and Operations Research*, North Holland, Amsterdam, 51–106.

B.S. Baker, E.G. Coffman Jr. (1981). A tight asymptotic bound for next–fit–decreasing bin packing. *SIAM Journal on Algebraic and Discrete Methods* **2**, 147–152.

E. Balas (1967). Discrete programming by the filter method. *Operations Research* **15**, 915–957.

E. Balas (1975). Facets of the knapsack polytope. *Mathematical Programming* **8**, 146–164.

E. Balas, R. Jeroslow (1972). Canonical cuts on the unit hypercube. *SIAM Journal of Applied Mathematics* **23**, 61–69.

E. Balas, R. Nauss, E. Zemel (1987). Comment on 'some computational results on real 0-1 knapsack problems'. *Operations Research Letters* **6**, 139.

E. Balas, E. Zemel (1978). Facets of the knapsack polytope from minimal covers. *SIAM Journal of Applied Mathematics* **34**, 119–148.

E. Balas, E. Zemel (1980). An algorithm for large zero-one knapsack problems. *Operations Research* **28**, 1130–1154.

R.S. Barr, G.T. Ross (1975). A linked list data structure for a binary knapsack algorithm. Research Report CCS 232, Centre for Cybernetic Studies, University of Texas.

R. Bellman (1954). Some applications of the theory of dynamic programming—a review. *Operations Research* **2**, 275–288.

R. Bellman (1957). *Dynamic Programming*, Princeton University Press, Princeton, NJ.

R. Bellman, S.E. Dreyfus (1962). *Applied Dynamic Programming*, Princeton University Press, Princeton, NJ.

R.L. Bulfin, R.G. Parker, C.M. Shetty (1979). Computational results with a branch and bound algorithm for the general knapsack problem. *Naval Research Logistics Quarterly* **26**, 41–46.

A.V. Cabot (1970). An enumeration algorithm for knapsack problems. *Operations Research* **18**, 306–311.

G. Carpaneto, S. Martello, P. Toth (1988). Algorithms and codes for the assignment problem. In B. Simeone, P. Toth, G. Gallo, F. Maffioli, S. Pallottino (eds), *Fortran Codes for Network Optimization, Annals of Operations Research* **13**, 193–223.

L. Chalmet, L. Gelders (1977). Lagrange relaxation for a generalized assignment-type problem. In M. Roubens (ed.), *Advances in Operations Research*, North-Holland, Amsterdam, 103–109.

S.K. Chang, A. Gill (1970a). Algorithmic solution of the change-making problem. *Journal of ACM* **17**, 113–122.

S.K. Chang, A. Gill (1970b). Algorithm 397. An integer programming problem. *Communications of ACM* **13**, 620–621.

L. Chang, J.F. Korsh (1976). Canonical coin-changing and greedy solutions. *Journal of ACM* **23**, 418–422.

N. Christofides, A. Mingozzi, P. Toth (1979). Loading problems. In N. Christofides, A. Mingozzi, P. Toth, C. Sandi (eds), *Combinatorial Optimization*, Wiley, Chichester, 339–369.

V. Chvátal (1980). Hard knapsack problems. *Operations Research* **28**, 402–411.

E.G. Coffman Jr., M.R. Garey, D.S. Johnson (1984). Approximation algorithms for bin-packing—an updated survey. In G. Ausiello, M. Lucertini, P. Serafini (eds), *Algorithm Design for Computer System Design*, Springer, Vienna, 49–106.

J. Cord (1964). A method for allocating funds to investment projects when returns are subject to uncertainty. *Management Science* **10**, 335–341.

H. Crowder, E.L. Johnson, M.W. Padberg (1983). Solving large-scale zero-one linear programming problems. *Operations Research* **31**, 803–834.

G.B. Dantzig (1957). Discrete variable extremum problems. *Operations Research* **5**, 266–277.

A. De Maio, C. Roveda (1971). An all zero-one algorithm for a certain class of transportation problems. *Operations Research* **19**, 1406–1418.

R.S. Dembo, P.L. Hammer (1980). A reduction algorithm for knapsack problems. *Methods of Operations Research* **36**, 49–60.

B.L. Dietrich, L.F. Escudero (1989a). More coefficient reduction for knapsack-like constraints in 0-1 programs with variable upper bounds. IBM T.J. Watson Research Center. RC-14389, Yorktown Heights (NY).

B.L. Dietrich, L.F. Escudero (1989b). New procedures for preprocessing 0-1 models with knapsack-like constraints and conjunctive and/or disjunctive variable upper bounds. IBM T.J. Watson Research Center. RC-14572, Yorktown Heights (NY).

K. Dudzinski, S. Walukiewicz (1984a). Upper bounds for the 0-1 knapsack problem. Report MPD–10–49/84, Systems Research Institute, Warsaw.

K. Dudzinski, S. Walukiewicz (1984b). A fast algorithm for the linear multiple-choice knapsack problem. *Operations Research Letters* **3**, 205–209.

K. Dudzinski, S. Walukiewicz (1987). Exact methods for the knapsack problem and its generalizations. *European Journal of Operational Research* **28**, 3–21.

M.E. Dyer (1984). An $O(n)$ algorithm for the multiple-choice knapsack linear program. *Mathematical Programming* **29**, 57–63.

M.E. Dyer, N. Kayal, J. Walker (1984). A branch and bound algorithm for solving the multiple-choice knapsack problem. *Journal of Computational and Applied Mathematics* **11**, 231–249.

S. Eilon, N. Christofides (1971). The loading problem. *Management Science* **17**, 259–267.

B. Faaland (1973). Solution of the value-independent knapsack problem by partitioning. *Operations Research* **21**, 332–337.

D. Fayard, G. Plateau (1975). Resolution of the 0-1 knapsack problem: comparison of methods. *Mathematical Programming* **8**, 272–307.

D. Fayard, G. Plateau (1982). An algorithm for the solution of the 0-1 knapsack problem. *Computing* **28**, 269–287.

M. Fischetti (1986). Worst-case analysis of an approximation scheme for the subset-sum problem. *Operations Research Letters* **5**, 283–284.

M. Fischetti (1989). A new linear storage, polynomial time approximation scheme for the subset-sum problem. *Discrete Applied Mathematics* (to appear).

M. Fischetti, S. Martello (1988). A hybrid algorithm for finding the kth smallest of n elements in $O(n)$ time. In B. Simeone, P. Toth, G. Gallo, F. Maffioli, S. Pallottino (eds), *Fortran Codes for Network Optimization, Annals of Operations Research* **13**, 401–419.

M. Fischetti, P. Toth (1988). A new dominance procedure for combinatorial optimization problems. *Operations Research Letters* **7**, 181–187.

M.L. Fisher (1980). Worst-case analysis of heuristic algorithms. *Management Science* **26**, 1–17.

M.L. Fisher (1981). The Lagrangian relaxation method for solving integer programming problems. *Management Science* **27**, 1–18.

M.L. Fisher, R. Jaikumar, L.N. Van Wassenhove (1986). A multiplier adjustment method for the generalized assignment problem. *Management Science* **32**, 1095–1103.

J.C. Fisk , M.S. Hung (1979). A heuristic routine for solving large loading problems. *Naval Research Logistics Quarterly* **26**, 643–650.

A.M. Frieze (1986). On the Lagarias-Odlyzko algorithm for the subset sum problem. *SIAM Journal on Computing* **15**, 536–539.

M.R. Garey, D.S. Johnson (1975). Complexity results for multiprocessor scheduling under resource constraints. *SIAM Journal on Computing* **4**, 397–411.

M.R. Garey, D.S. Johnson (1978). "Strong" NP-completeness results: motivation, examples and implications. *Journal of ACM* **25**, 499–508.

M.R. Garey, D.S. Johnson (1979). *Computers and Intractability: a Guide to the Theory of NP-Completeness*, Freeman, San Francisco.

R.S. Garfinkel, G.L. Nemhauser (1972). *Integer Programming*, John Wiley and Sons, New York.

G.V. Gens, E.V. Levner (1978). Approximation algorithms for scheduling problems. *Izvestija Akademii Nauk SSSR, Engineering Cybernetics* **6**, 38–43.

G.V. Gens, E.V. Levner (1979). Computational complexity of approximation algorithms for combinatorial problems. In J. Bečvař (ed.), *Mathematical Foundations of Computer Science 1979*, Lecture Notes in Computer Science 74, Springer, Berlin, 292–300.

G.V. Gens, E.V. Levner (1980). Fast approximation algorithms for knapsack type problems. In K. Iracki, K. Malinowski, S. Walukiewicz (eds), *Optimization Techniques, Part 2*, Lecture Notes in Control and Information Sciences 23, Springer, Berlin, 185–194.

A. Geoffrion (1969). An improved implicit enumeration approach for integer programming. *Operations Research* **17**, 437–454.

P.C. Gilmore, R.E. Gomory (1961). A linear programming approach to the cutting stock problem I. *Operations Research* **9**, 849–858.

P.C. Gilmore, R.E. Gomory (1963). A linear programming approach to the cutting stock problem II. *Operations Research* **11**, 863–888.

P.C. Gilmore, R.E. Gomory (1965). Multi-stage cutting stock problems of two and more dimensions. *Operations Research* **13**, 94–120.

P.C. Gilmore, R.E. Gomory (1966). The theory and computation of knapsack functions. *Operations Research* **14**, 1045–1074.

F. Glover (1965). A multiphase dual algorithm for the zero-one integer programming problem. *Operations Research* **13**, 879–919.

F. Glover, D. Klingman (1979). A $o(n \log n)$ algorithm for LP knapsacks with GUB constraints. *Mathematical Programming* **17**, 345–361.

A.V. Goldberg, A. Marchetti-Spaccamela (1984). On finding the exact solution to a zero-one knapsack problem. *Proc. 16th Annual ACM Symposium Theory of Computing*, 359–368.

E.S. Gottlieb, M.R. Rao (1988). Facets of the knapsack polytope derived from disjoint and overlapping index configurations. *Operations Research Letters* **7**, 95–100.

E.S. Gottlieb, M.R. Rao (1989a). The generalized assignment problem: valid inequalities and facets. *Mathematical Programming* (to appear).

E.S. Gottlieb, M.R. Rao (1989b). (1,k)-configuration facets for the generalized assignment problem. *Mathematical Programming* (to appear).

H. Greenberg (1985). An algorithm for the periodic solutions in the knapsack problem. *Journal of Mathematical Analysis and Applications* **111**, 327–331.

H. Greenberg (1986). On equivalent knapsack problems. *Discrete Applied Mathematics* **14**, 263–268.

H. Greenberg, I. Feldman (1980). A better-step-off algorithm for the knapsack problem. *Discrete Applied Mathematics* **2**, 21–25.

H. Greenberg, R.L. Hegerich (1970). A branch search algorithm for the knapsack problem. *Management Science* **16**, 327–332.

M.M. Guignard, S. Kim (1987). Lagrangean decomposition: A model yielding stronger Lagrangean bounds. *Mathematical Programming* **39**, 215–228.

M.M. Guignard, K. Spielberg (1972). Mixed-integer algorithms for the (0,1) knapsack problem. *IBM Journal of Research and Development* **16**, 424–430.

P.L. Hammer, E.L. Johnson, U.N. Peled (1975). Facets of regular 0-1 polytopes. *Mathematical Programming* **8**, 179–206.

D. Hartvigsen, E. Zemel (1987). On the complexity of lifted inequalities for the knapsack problem. Report 740, Department of Managerial Economics and Decision Sciences, Northwestern University, Evanston, Illinois.

D.S. Hirschberg, C.K. Wong (1976). A polynomial-time algorithm for the knapsack problem with two variables. *Journal of ACM* **23**, 147–154.

E. Horowitz, S. Sahni (1974). Computing partitions with applications to the knapsack problem. *Journal of ACM* **21**, 277–292.

T.C. Hu (1969). *Integer Programming and Network Flows*, Addison-Wesley, New York.

T.C. Hu, M.L. Lenard (1976). Optimality of a heuristic solution for a class of knapsack problems. *Operations Research* **24**, 193–196.

P.D. Hudson (1977). Improving the branch and bound algorithms for the knapsack problem. Queen's University Research Report, Belfast.

M.S. Hung, J.R. Brown (1978). An algorithm for a class of loading problems. *Naval Research Logistics Quarterly* **25**, 289–297.

M.S. Hung, J.C. Fisk (1978). An algorithm for 0-1 multiple knapsack problems. *Naval Research Logistics Quarterly* **24**, 571–579.

O.H. Ibarra, C.E. Kim (1975). Fast approximation algorithms for the knapsack and sum of subset problems. *Journal of ACM* **22**, 463–468.

G.P. Ingargiola, J.F. Korsh (1973). A reduction algorithm for zero-one single knapsack problems. *Management Science* **20**, 460–463.

G.P. Ingargiola, J.F. Korsh (1975). An algorithm for the solution of 0-1 loading problems. *Operations Research* **23**, 1110–1119.

G.P. Ingargiola, J.F. Korsh (1977). A general algorithm for one-dimensional knapsack problems. *Operations Research* **25**, 752–759.

D.S. Johnson (1973). Near-optimal bin packing algorithms. Technical Report MAC TR-109, Project MAC, Massachusetts Institute of Technology, Cambridge, MA.

D.S. Johnson (1974). Approximation algorithms for combinatorial problems. *Journal of Computer and System Sciences* **9**, 256–278.

D.S. Johnson, A. Demers, J.D. Ullman, M.R. Garey, R.L. Graham (1974). Worst-case performance bounds for simple one-dimensional packing algorithms. *SIAM Journal on Computing* **3**, 299–325.

S.C. Johnson, B.W. Kernighan (1972). Remarks on algorithm 397. *Communications of ACM* **15**, 469.

K. Jörnsten, M. Näsberg (1986). A new Lagrangian relaxation approach to the generalized assignment problem. *European Journal of Operational Research* **27**, 313–323.

R. Kannan (1980). A polynomial algorithm for the two-variables integer programming problem. *Journal of ACM* **27**, 118–122.

S. Kaplan (1966). Solution of the Lorie-Savage and similar integer programming problems by the generalized Lagrange multiplier method. *Operations Research* **14**, 1130–1136.

R.M. Karp (1972). Reducibility among combinatorial problems. In R.E. Miller, J.W. Thatcher (eds), *Complexity of Computer Computations*, Plenum Press, New York, 85–103.

R.M. Karp, J.K. Lenstra, C.J.H. McDiarmid, A.H.G. Rinnooy Kan (1985). Probabilistic analysis. In M. O'hEigeartaigh, J.K. Lenstra, A.H.G. Rinnooy Kan (eds), *Combinatorial Optimization: Annotated Bibliographies*, Wiley, Chichester, 52–88.

T.D. Klastorin (1979). An effective subgradient algorithm for the generalized assignment problem. *Computers and Operations Research* **6**, 155–164.

D.E. Knuth (1973). *The Art of Computer Programming, Vol. 3, Sorting and Searching*, Addison-Wesley, Reading, MA.

P.J. Kolesar (1967). A branch and bound algorithm for the knapsack problem. *Management Science* **13**, 723–735.

N.W. Kuhn (1955). The Hungarian method for the assignment problem. *Naval Research Logistics Quarterly* **2**, 83–97.

J.C. Lagarias, A.M. Odlyzko (1983). Solving low-density subset sum problems. *Proc. 24th Annual IEEE Symposium Foundations of Computer Science*, 1–10.

B.J. Lageweg, J.K. Lenstra (1972). Algoritmend voor knapzack problemen. Report BN 14/72, Stichting Mathematisch Centrum, Amsterdam.

M. Laurière (1978). An algorithm for the 0-1 knapsack problem. *Mathematical Programming* **14**, 1–10.

E.L. Lawler (1976). *Combinatorial Optimization: Networks and Matroids*, Holt, Rinehart & Winston, New York.

E.L. Lawler (1979). Fast approximation algorithms for knapsack problems. *Mathematics of Operations Research* **4**, 339–356.

E.V. Levner, G.V. Gens (1978). *Discrete Optimization Problems and Approximation Algorithms*. Moscow, CEMI (Russian).

G.S. Lueker (1975). Two NP-complete problems in nonnegative integer programming. Report No. 178, Computer Science Laboratory, Princeton University, Princeton, NJ.

G.S. Lueker (1982). On the average difference between the solutions to linear and integer knapsack problems. In R.L. Disney, T.J. Ott (eds), *Applied Probability—Computer Science: the Interface, Vol. I*, Birkhauser, Basel, 489–504.

N. Maculan (1983). Relaxation Lagrangienne: le problème du knapsack 0-1. INFOR *(Canadian Journal of Operational Research and Information Processing)* **21**, 315–327.

M.J. Magazine, J.L. Nemhauser, L.E. Trotter Jr. (1975). When the greedy solution solves a class of knapsack problems. *Operations Research* **23**, 207–217.

M.J. Magazine, O. Oguz (1981). A fully polynomial approximate algorithm for the 0-1 knapsack problem. *European Journal of Operational Research* **8**, 270–273.

A. Marchetti-Spaccamela, C. Vercellis (1987). Efficient on-line algorithms for the knapsack problem. In T. Ottman (ed.), *Automata, Languages and Programming*, Lecture Notes in Computer Science 267, Springer, Berlin, 445–456.

S. Martello, P. Toth (1977a). An upper bound for the zero-one knapsack problem and a branch and bound algorithm. *European Journal of Operational Research* **1**, 169–175.

S. Martello, P. Toth (1977b). Computational experiences with large-size unidimensional knapsack problems. Presented at the TIMS/ORSA Joint National Meeting, San Francisco.

S. Martello, P. Toth (1977c). Solution of the bounded and unbounded change-making problem. Presented at the TIMS/ORSA Joint National Meeting, San Francisco.

S. Martello, P. Toth (1977d). Branch and bound algorithms for the solution of the general unidimensional knapsack problem. In M. Roubens (ed.), *Advances in Operations Research*, North-Holland, Amsterdam, 295–301.

S. Martello, P. Toth (1978). Algorithm for the solution of the 0-1 single knapsack problem. *Computing* **21**, 81–86.

S. Martello, P. Toth (1979). The 0-1 knapsack problem. In N. Christofides, A. Mingozzi, P. Toth, C. Sandi (eds), *Combinatorial Optimization*, Wiley, Chichester, 237–279.

S. Martello, P. Toth (1980a). Solution of the zero-one multiple knapsack problem. *European Journal of Operational Research* **4**, 276–283.

S. Martello, P. Toth (1980b). Optimal and canonical solutions of the change-making problem. *European Journal of Operational Research* **4**, 322–329.

S. Martello, P. Toth (1980c). A note on the Ingargiola–Korsh algorithm for one-dimensional knapsack problems. *Operations Research* **28**, 1226–1227.

S. Martello, P. Toth (1981a). A bound and bound algorithm for the zero-one multiple knapsack problem. *Discrete Applied Mathematics* **3**, 275–288.

S. Martello, P. Toth (1981b). Heuristic algorithms for the multiple knapsack problem. *Computing* **27**, 93–112.

S. Martello, P. Toth (1981c). An algorithm for the generalized assignment problem. In J.P. Brans (ed.), *Operational Research '81*, North-Holland, Amsterdam, 589–603.

S. Martello, P. Toth (1984a). A mixture of dynamic programming and branch-and-bound for the subset-sum problem. *Management Science* **30**, 765–771.

S. Martello, P. Toth (1984b). Worst-case analysis of greedy algorithms for the subset-sum problem. *Mathematical Programming* **28**, 198–205.

S. Martello, P. Toth (1985a). Approximation schemes for the subset-sum problem: survey and experimental analysis. *European Journal of Operational Research* **22**, 56–69.

S. Martello, P. Toth (1985b). Algorithm 632. A program for the 0-1 multiple knapsack problem. *ACM Transactions on Mathematical Software* **11**, 135–140.

S. Martello, P. Toth (1987). Algorithms for knapsack problems. In S. Martello, G. Laporte, M. Minoux, C. Ribeiro (eds), *Surveys in Combinatorial Optimization, Annals of Discrete Mathematics* 31, North-Holland, Amsterdam, 213–257.

S. Martello, P. Toth (1988). A new algorithm for the 0-1 knapsack problem. *Management Science* **34**, 633–644.

S. Martello, P. Toth (1989). An exact algorithm for the bin packing problem. Presented at EURO X, Beograd.

S. Martello, P. Toth (1990a). An exact algorithm for large unbounded knapsack problems. *Operations Research Letters* (to appear).

S. Martello, P. Toth (1990b). Lower bounds and reduction procedures for the bin packing problem. *Discrete Applied Mathematics* (to appear).

J.B. Mazzola (1989). Generalized assignment with nonlinear capacity interaction. *Management Science* **35**, 923–941.

M. Meanti, A.H.G. Rinnooy Kan, L. Stougie, C. Vercellis (1989). A probabilistic analysis of the multiknapsack value function. *Mathematical Programming* (to appear).

H. Müller-Merbach (1978). An improved upper bound for the zero-one knapsack problem: a note on the paper by Martello and Toth. *European Journal of Operational Research* **2**, 212–213.

R.A. Murphy (1986). Some computational results on real 0-1 knapsack problems. *Operations Research Letters* **5**, 67–71.

R.M. Nauss (1976). An efficient algorithm for the 0-1 knapsack problem. *Management Science* **23**, 27–31.

R.M. Nauss (1978). The 0-1 knapsack problem with multiple choice constraints. *European Journal of Operational Research* **2**, 125–131.

A. Neebe, D. Dannenbring (1977). Algorithms for a specialized segregated storage problem. Technical Report 77–5, University of North Carolina.

G.L. Nemhauser, L.E. Trotter (1974). Properties of vertex packing and independence system polyhedra. *Mathematical Programming* **6**, 48–61.

G.L. Nemhauser, Z. Ullmann (1969). Discrete dynamic programming and capital allocation. *Management Science* **15**, 494–505.

G.L. Nemhauser, L.A. Wolsey (1988). *Integer and Combinatorial Optimization*, Wiley, Chichester.

M.W. Padberg (1975). A note on zero-one programming. *Operations Research* **23**, 833–837.

M.W. Padberg (1979). Covering, packing and knapsack problems. *Annals of Discrete Mathematics* **4**, 265–287.

M.W. Padberg (1980). (1,k)-configurations and facets for packing problems. *Mathematical Programming* **18**, 94–99.

C.H. Papadimitriou, K. Steiglitz (1982). *Combinatorial Optimization*, Prentice-Hall, Englewood Cliffs, NJ.

G. Plateau, M. Elkihel (1985). A hybrid algorithm for the 0-1 knapsack problem. *Methods of Operations Research* **49**, 277–293.

W.R. Pulleyblank (1983). Polyhedral combinatorics. In A. Bachem, M. Grötschel, B. Korte (eds), *Mathematical Programming: the State of the Art–Bonn 1982*, Springer, Berlin, 312–345.

A.H.G. Rinnooy Kan (1987). Probabilistic analysis of algorithms. In S. Martello, G. Laporte, M. Minoux, C. Ribeiro (eds), *Surveys in Combinatorial Optimization, Annals of Discrete Mathematics* **31**, North-Holland, Amsterdam, 365–384.

G.T. Ross, R.M. Soland (1975). A branch and bound algorithm for the generalized assignment problem. *Mathematical Programming* **8**, 91–103.

B.F. Ryder, A.D. Hall (1981). The PFORT verifier. Computer Science Report 2, Bell Laboratories.

S. Sahni (1975). Approximate algorithms for the 0-1 knapsack problem. *Journal of ACM* **22**, 115–124.

S. Sahni, T. Gonzalez (1976). P-complete approximation problems. *Journal of ACM* **23**, 555–565.

H.M. Salkin (1975). *Integer Programming*, Addison-Wesley, New York.

H.M. Salkin, C.A. de Kluyver (1975). The knapsack problem: a survey. *Naval Research Logistics Quarterly* **22**, 127–144.

A. Schrijver (1986). *Theory of Linear and Integer Programming*, Wiley, Chichester.

P. Sinha, A.A. Zoltners (1979). The multiple-choice knapsack problem. *Operations Research* **27**, 503–515.

V. Srinivasan, G.L. Thompson (1973). An algorithm for assigning uses to sources in a special class of transportation problems. *Operations Research* **21**, 284–295.

U. Suhl (1978). An algorithm and efficient data structures for the binary knapsack problem. *European Journal of Operational Research* **2**, 420–428.

M.M. Syslo, N. Deo, J.S. Kowalik (1983). *Discrete Optimization Algorithms with Pascal Programs*, Prentice-Hall, Englewood Cliffs, NJ.

K. Szkatula, M. Libura (1987). On probabilistic properties of greedy-like algorithms for the binary knapsack problem. Report 154, Instytut Badan Systemowych, Polska Akademia Nauk, Warsaw.

H.A. Taha (1975). *Integer Programming*, Academic Press, New York.

B.N. Tien, T.C. Hu (1977). Error bounds and the applicability of the greedy solution to the coin-changing problem. *Operations Research* **25**, 404–418.

G. Tinhofer, H. Schreck (1986). The bounded subset sum problem is almost everywhere randomly decidable in $O(n)$. *Information Processing Letters* **23**, 11–17.

M. Todd (1980). Theorem 3. In V. Chvátal. Hard knapsack problems, *Operations Research* **28**, 1408–1409.

P. Toth (1976). A new reduction algorithm for 0-1 knapsack problems. Presented at the ORSA/TIMS Joint National Meeting, Miami.

P. Toth (1980). Dynamic programming algorithms for the zero-one knapsack problem. *Computing* **25**, 29–45.

G.P. Veliev, K.Sh. Mamedov (1981). A method of solving the knapsack problem. *USSR Computational Mathematics and Mathematical Physics* **21**, 75–81.

A. Verebriusova (1904). On the number of solutions of indefinite equations of the first degree with many variables. *Mathematicheskii Sbornik* **24**, 662–688.

P.R.C. Villela, C.T. Bornstein (1983). An improved bound for the 0-1 knapsack problem. Report ES31-83, COPPE-Federal University of Rio de Janeiro.

H.M. Weingartner (1963). *Mathematical Programming and the Analysis of Capital Budgeting Problems*, Prentice-Hall, Englewood Cliffs, NJ.

H.M. Weingartner (1968). Capital budgeting and interrelated projects: survey and synthesis. *Management Science* **12**, 485–516.

H.M. Weingartner, D.N. Ness (1967). Methods for the solution of the multi-dimensional 0-1 knapsack problem. *Operations Research* **15**, 83–103.

L.A. Wolsey (1975). Faces of linear inequalities in 0-1 variables. *Mathematical Programming* **8**, 165–178.

J.W. Wright (1975). The change-making problem. *Journal of ACM* **22**, 125–128.

E. Zemel (1978). Lifting the facets of zero-one polytopes. *Mathematical Programming* **15**, 268–277.

E. Zemel (1980). The linear multiple choice knapsack problem. *Operations Research* **28**, 1412–1423.

E. Zemel (1984). An $O(n)$ algorithm for the linear multiple choice knapsack problem and related problems. *Information Processing Letters* **18**, 123–128.

E. Zemel (1988). Easily computable facets of the knapsack polytope. Report 713, Department of Managerial Economics and Decision Sciences, Northwestern University, Evanston, Illinois.

A.A. Zoltners (1978). A direct descent binary knapsack algorithm. *Journal of ACM* **25**, 304–311.

Author index

Note: listing in references section is indicated by bold page numbers.

Subject index

Note: abbreviations used in the text and in this index:

BCMP	= Bounded Change-Making Problem
BKP	= Bounded Knapsack Problem
BPP	= Bin-Packing Problem
CMP	= Change-Making Problem
GAP	= Generalized Assignment Problem
KP	= 0-1 Knapsack Problem
MCKP	= Multiple-Choice Knapsack Problem
MKP	= 0-1 Multiple Knapsack Problem
SSP	= Subset-Sum Problem
UEMKP	= Unbounded Equality Constrained Min-Knapsack Problem
UKP	= Unbounded Knapsack Problem